T0245235

CAMBRIDGE LIBRARY COLLECTION

Books of enduring scholarly value

Earth Sciences

In the nineteenth century, geology emerged as a distinct academic discipline. It pointed the way towards the theory of evolution, as scientists including Gideon Mantell, Adam Sedgwick, Charles Lyell and Roderick Murchison began to use the evidence of minerals, rock formations and fossils to demonstrate that the earth was older by millions of years than the conventional, Bible-based wisdom had supposed. They argued convincingly that the climate, flora and fauna of the distant past could be deduced from geological evidence. Volcanic activity, the formation of mountains, and the action of glaciers and rivers, tides and ocean currents also became better understood. This series includes landmark publications by pioneers of the modern earth sciences, who advanced the scientific understanding of our planet and the processes by which it is constantly re-shaped.

The Life and Correspondence of William Buckland, D.D., F.R.S.

First published in 1894, this biography details the life of renowned geologist William Buckland (1784–1856) who, along with Sedgwick and Lyell, was one of the pioneers of modern geological inquiry. While he is better known for attempting to correlate his geological findings with the Old Testament, Buckland's studies paved the way for Darwin's development of evolutionary theory. In the course of his illustrious career, Buckland was a Canon of Christ Church, was twice appointed President of the Geological Society, served as the first President of the British Association, and became Dean of Westminster. Penned by Buckland's daughter almost forty years after his death, *The Life and Correspondence* provides a more personal insight into Buckland's scientific endeavours. Gordon's biography is complemented by several illustrations, and the appendices include an extensive list of positions held by Buckland and his membership of learned societies, and a complete index of his publications.

Cambridge University Press has long been a pioneer in the reissuing of out-of-print titles from its own backlist, producing digital reprints of books that are still sought after by scholars and students but could not be reprinted economically using traditional technology. The Cambridge Library Collection extends this activity to a wider range of books which are still of importance to researchers and professionals, either for the source material they contain, or as landmarks in the history of their academic discipline.

Drawing from the world-renowned collections in the Cambridge University Library, and guided by the advice of experts in each subject area, Cambridge University Press is using state-of-the-art scanning machines in its own Printing House to capture the content of each book selected for inclusion. The files are processed to give a consistently clear, crisp image, and the books finished to the high quality standard for which the Press is recognised around the world. The latest print-on-demand technology ensures that the books will remain available indefinitely, and that orders for single or multiple copies can quickly be supplied.

The Cambridge Library Collection will bring back to life books of enduring scholarly value (including out-of-copyright works originally issued by other publishers) across a wide range of disciplines in the humanities and social sciences and in science and technology.

The Life and Correspondence of William Buckland, D.D., F.R.S.

*Sometime Dean of Westminster, Twice
President of the Geological Society, and First
President of the British Association*

ELIZABETH OKE GORDON

CAMBRIDGE
UNIVERSITY PRESS

CAMBRIDGE UNIVERSITY PRESS

Cambridge, New York, Melbourne, Madrid, Cape Town, Singapore,
São Paolo, Delhi, Dubai, Tokyo

Published in the United States of America by Cambridge University Press, New York

www.cambridge.org
Information on this title: www.cambridge.org/9781108021630

This edition first published 1894
This digitally printed version 2010

ISBN 978-1-108-02163-0 Paperback

THE

LIFE AND CORRESPONDENCE

OF

WILLIAM BUCKLAND, D.D., F.R.S.

Thompson, Pinx   Walker & Boutall. Ph. Sc.

ætat 62

THE

LIFE AND CORRESPONDENCE

OF

WILLIAM BUCKLAND, D.D., F.R.S.,

SOMETIME DEAN OF WESTMINSTER, TWICE PRESIDENT OF
THE GEOLOGICAL SOCIETY, AND FIRST PRESIDENT
OF THE BRITISH ASSOCIATION.

BY HIS DAUGHTER,

MRS. GORDON.

"Out of the old fieldes, as men saith,
   Cometh all this new corn fro' year to year;
And out of old bookes, in good faith,
   Cometh all this new science that men lere."

Chaucer.

WITH PORTRAITS AND ILLUSTRATIONS.

LONDON:

JOHN MURRAY, ALBEMARLE STREET.

1894.

PREFACE.

THE century now drawing to a close is remarkable beyond all others for the spirit of inquiry into the physical constitution of the earth, into the forces playing upon its surface, and into the phenomena of life, both plant and animal. The rise of the natural sciences in the modern sense may be said to date from its beginning. In this great renascence geology has borne an important part. It has opened out new and almost endless avenues of thought, giving us, on the one hand, the history of the ever-changing earth, from the remote time when it was sufficiently cool to allow of water resting upon its surface, and, on the other, the long and orderly procession of animal life beginning with the lowest invertebrate forms and ending in Man. In this latter connection it enabled Darwin to grasp the principle of evolution that now influences our view of life as a whole in the same way as the law of gravitation has affected our view of matter, not only in the earth, but also in the universe. To us, living at the end of the century, it is difficult to realise the conditions under which the pioneers lived and worked, because through their labours the conditions have wholly changed. In this short

Life of Dr. Buckland, written under considerable difficulty and nearly four decades after his death, we are brought face to face with the old order of things, and we can realise how great is the evolution that has taken place since his time. It is a sketch of no mere personal interest, but valuable as throwing light upon social and scientific conditions which have long passed away. It illustrates the position of science at Oxford during the first fifty years of the century.

It also fills a blank in the history of the founders of geology—William Smith, Sedgwick, De la Bêche, Murchison, Phillips, and Lyell. Among these Buckland stands in the foremost rank. He began his work earlier than any of them, excepting William Smith, and the main difference between him and Sedgwick lies in the fact that he was a geologist from his youth up, while Sedgwick, strangely enough, was allured into geological studies by being appointed Woodwardian Professor at Cambridge.[1]

In this preface, made at the request of the authoress, I shall draw attention to those points in Buckland's geological career which appear to me, an Oxford man long after his time, and profoundly influenced by his work, to be most noteworthy. Of the other aspects of his many-sided genius I shall say nothing. Nor shall I say anything about his advancement in the Church or of his social position at

---

[1] This statement sounds almost incredible. We have it, however, on Sedgwick's own authority. On his appointment, he said characteristically: "Hitherto I have never turned a stone, now I will leave no stone unturned." His friend Dr. Ainger, congratulating him on the appointment, writes that it "will sometimes lead you to pile up stones, as well as to range them in your lecture-room."

Westminster in his later years. Most men cease to be interesting after they have gained their success in life. Buckland was full of interest to the end.

Buckland graduated with distinction at Corpus Christi College, Oxford, in 1804, in the golden days long before honours and class-lists were dreamt of. Five years later he was ordained and elected Fellow. As a boy he had taken a keen interest in the rocks and fossils of his Devonshire home, and at Winchester, where he was at school, and, early on his arrival at Oxford, had fallen under the influence of William Smith, "the Father of English Geology." He took his first lesson in field geology from one of William Smith's friends. The fruits of this walk to Shotover formed the nucleus of the collection which ultimately expanded into the present Geological Museum. There was in those days nothing of the nature of a Museum in Oxford, excepting the miscellaneous collection of curiosities and antiquities founded by Elias Ashmole. Buckland turned his rooms into a museum, and Murchison has graphically described him sitting in the only empty chair, in his black gown, cleaning out a fossil bone from its matrix, and surrounded by rocks, shells, and bones in dire confusion. Academical dress, it must be noted, was then worn in walks into the country even as far as Shotover.

In 1813 he was appointed Reader in Mineralogy, and his influence as a lecturer was so strongly felt that five years later the Readership of Geology was created for him in the University, in the very year when Sedgwick was appointed to the old-established Woodwardian Professorship at Cambridge. His wit, humour, and eloquence attracted both young and old, and the memory of his geological expeditions had not perished when I was an undergraduate in 1859. His rooms at Christ Church, to which he had migrated on his appointment as Canon, became a centre of attraction for all who cared for the new learning that by this time was grievously vexing the minds of mediæval Oxford. How strong was the feeling of antagonism, even after many years, may be estimated by the pious ejaculation of Dean Gaisford in 1852: "Buckland has gone to Italy, and we shall hear no more, thank God, of this geology!" They were, however, to hear more, both of this and of other things too, until the spirit of narrow intolerance received its crushing defeat in the memorable Darwinian controversy in 1860. In this widening of thought, and in sweeping away the old worn-out ideas of Nature, Buckland did most important service to the University. Single-handed, he brought about a revival in the direction of natural science, analogous to the movement in religious thought brought about by Newman and the Oriel School.

The phrase " gnoscitur e sociis " applies to all men, but with peculiar force to a professor. Buckland was in close touch with the most brilliant men of the day and of most varied pursuits. Whately, Whewell, Sir Robert Peel, Cuvier, Humboldt, Liebig, and Sir Joseph Banks were among his friends. It is, however, by his influence on his students that he can best be measured. Among these, two young Christ Church men may be mentioned—Viscount Cole, afterwards the Earl of Enniskillen, and Philip Egerton, afterwards the baronet. Both these men moulded their lives on his teaching, and enriched geological science by their papers and collections. Among his students now living, Sir Henry Acland, Storey Maskelyne, and Ruskin have borne witness in these pages to his power. He was the founder of the new learning in Oxford, and he started the movement which has borne fruit in the present place of the natural sciences in the studies of the University.

Buckland's influence, however, was felt as a teacher and master far beyond Oxford. To him Murchison owed his first lesson in the field, and his first " true launch " in 1824 into the line of work in which he was in after years to do so much. To him, in 1831, Murchison turned for advice and assistance when he had decided to attack the difficult problem of Welsh geology, and from him he obtained the clue to the true sequence of the rocks below the Old Red Sandstone on the banks of the Wye that led ultimately to the Silurian System. To him, too, is due the discovery of the value of the phosphates in the coprolite beds that has contributed so much to the development of modern agriculture. In this connection Lord Playfair bears ample witness, and tells us in this Life how much he owes to Buckland's friendship and guidance.

When the history of the progress of geological knowledge comes to be written, the work of the Geological Society of London in organising and directing individual effort will be fully recognised. Founded in 1807 by Grenough, it attracted some of the acutest intellects of the day— Wollaston, Warburton, Fitton, and others. Buckland joined it in 1817, and Sedgwick in the following year.

In 1824, when it was formally incorporated by charter, Buckland became its President. It was composed "of robust, joyous, and independent spirits, who toiled well in the field, and did battle and cuffed opinions with much spirit and great good will." Murchison and Lyell were among the younger members. Buckland took a leading share in the debates of the Society and in contributing papers down to the middle of the century. He was one of the first to recognise the existence of glaciers in this country, and wrote a paper in 1840 on their evidences in Scotland and in the north of England. In the debate he was vigorously opposed by Murchison and Whewell, and equally vigorously supported by Lyell and Agassiz. Buckland in reply summed up the arguments, and condemned all who dared to doubt the orthodoxy of the grooves and scratches of the ice-worn mountains to "suffer the pains of eternal itch without the privilege of scratching." This characteristic debate, following papers by Agassiz and Buckland, marks the beginning of the glacial controversy, which has divided geological opinion ever since.

Buckland also was one of the founders of the British Association, and was the first President after its formal organisation at Oxford in 1832.[1] It was this meeting which made the Association an assured success. It is no small testimony to the high place of geology among the sciences at this time, that Sedgwick should have succeeded to the presidential chair in the following year at Cambridge.

In the first thirty years of the century the Diluvial

---

[1] The first meeting was at York in the previous year, which Buckland was unable to attend.

theory, or, in other words, the Noachian deluge, was held to be a sufficient explanation of the sand, gravel, and clay containing marine and freshwater shells, and the bones of mammalia, which lie scattered over wide areas on the land and occur also in the ossiferous caverns. With this idea in his mind Buckland explored in 1821 the bone cave at Kirkdale, and recorded the general results of his examination of caves and of the diluvium in Britain and on the Continent in the "Reliquiæ Diluvianæ." While he accepted the general evidence as to the Noachian deluge, he fully recognised that the Kirkdale cave was a den of hyenas, and that they had dragged in the other animals found in it for food. This book led to the more minute study of bone caverns, and ultimately to the wonderful discoveries in the caves of this country and of the Continent, which have revealed to us the existence of man hunting the reindeer, musk-sheep, and mammoth in France, Germany, and Britain, and living at a time when all the animals found in Kirkdale could wander freely northwards and westwards from the Alps and Pyrenees to the coast now marked by the hundred-fathom line in the Atlantic. It was this work that led me in 1859 into the path of comparative osteology, and to the exploration of Wookey Hole and other ossiferous caverns.

The "Reliquiæ Diluvianæ" still remains the best book on caves. Buckland, it must be remarked, gave up the diluvial theory, as he came to recognise the power of ice, and the truth of the uniformitarian doctrine of the operation of existing causes in past times. It is not a little strange that it should have been revived by Prestwich, his successor at Oxford, and by Howorth, some fifty years afterwards.

In concluding these remarks I would remind the reader that Buckland belongs to the heroic age, when Natural Science was young, and that he belongs to a type of man now extinct. Whatever estimate may be formed of his life and works, it cannot be denied that he was one of the makers of modern Oxford, and one of the founders of the science of Geology.

W. BOYD DAWKINS.

OWENS COLLEGE, MANCHESTER,
September 22nd, 1894.

CONTENTS.

CHAPTER I.

1784—1808.

CHAPTER II.

1808—1822.

CHAPTER III.

1822—1824.

CHAPTER IV.

1825—1830.

CHAPTER V.

1831—1841.

CHAPTER VI.

1839—1845.

CHAPTER X.

APPENDIX.

LIST OF ILLUSTRATIONS.

LIFE OF DEAN BUCKLAND.

CHAPTER I.

WILLIAM BUCKLAND was the eldest son of the Rev. Charles Buckland, Rector of Templeton and Trusham in the county of Devon. He was born at Axminster, on the 12th of March, 1784. His mother, whose maiden name was Elizabeth Oke, was the daughter of Mr. John Oke, a landed proprietor, living at Combpyne, near Axminster, whose family had since the Stuart period occupied extensive property in that neighbourhood.

The birthplace of William Buckland was singularly adapted to develope his peculiar genius. Near his home, in the picturesque valley of the Axe, are large quarries of lias, abounding· in fossil organic remains ; in this same valley are also found abundant traces of a buried forest ; here, too, lay embedded among the roots of the trees the bones of fossil elephants. His father (who for the last twenty-two years of his life was blind from an accident) early made his son the companion of his walks and tastes. Together they ransacked the

I

lias quarries, collecting ammonites and other shells, which thus became familiar to the lad from his infancy. From his, childhood his innate faculties for observation were encouraged. Writing of this early period of his life to the late Sir H. de la Bêche, Dr. Buckland himself says : " The love of observing natural objects which is common to most children was early exhibited by my aptitude in finding birds' nests and collecting their eggs. I also made observations on the habits of fishes in the Axe—particularly flounders, minnows, roaches, eels, and miller's thumbs."

One of William Buckland's earliest and most intimate companions, the late William Daniel Conybeare (afterwards Dean of Llandaff), has noticed the peculiar concurrence of circumstances which fostered the natural genius of the boy. In the following extract from a letter written to Frank Buckland, Mr. Conybeare speaks of his friend's youthful days :—

" All the circumstances of Buckland's early life were calculated to impress that character of mind which so peculiarly qualified him to become the pioneer of the rising science of Geology, which began to unfold itself during the very period when his powers first acquired mature development. Those powers were, from a child, marked by an eager curiosity of investigation, and by resolute and unwearied activity of observation and research ; anything at all novel and striking at once attracted his eye, and he was discontented until he had succeeded in tracing out all the dependencies connected with the objects which attracted him, and had thus fully made out and illustrated their history.

" The very place of his birth itself did much to call his

early attention to the marvels of the fossil remains of the organised beings which had occupied our planet in its earlier stages of progress, and the various strata of its coast in which these singular relics lie embedded. The town of Axminster, on the confines of Devon and Dorset, is situated in a valley based on that peculiar rock formation, the lias, which is most rich in organic remains, and exhibits so many of their most striking and interesting forms. Axminster is within a few miles of the most illustrative of those coast sections exposing the structure and contents of that rock, and its connexion with various other overlying secondary deposits of oolitic and cretaceous rocks and underlying masses of the new red sandstone. All these features are brought so prominently forward and exhibited in so close a compass, that a child of sagacity, growing up among them, could hardly fail to have its mind impressed with the elements of practical geology, though as yet ignorant of the science.

" The young Buckland could not take a stroll in the neighbouring fields without stumbling, at almost every step, on lias quarries, and finding, on ascending every hill, that its summit consisted of an entirely dissimilar formation—chertsand. If he extended his rambles to the shore at Lyme Regis or Charmouth, crowds of little urchins ran after him to tempt him with pretty little golden serpents (pyritous ammonites) or wonderful thunderbolts (belemnites), and he must soon have learnt to find for himself the situations in which these treasures abounded. He must have found himself able to walk for miles over the slabs which the lias protruded into the sea, without placing a foot beyond the numerous circles of the larger varieties of his serpent-stones, and found the supposed belemnites aggregated in thousands in particular portions of the cliff. If therefore, turning home, he sauntered over Lyme Cobb, his eye must have been caught by the rich and variously coloured panorama of the coast section before him. We seldom find a child brought up near the sea as ignorant as an inland child ; his little box of treasures generally is filled with various shells and marine curiosities, and he readily learns to discriminate the peculiarity of their forms ; and if he has any curiosity he will naturally be led to speculate on the uses of their several parts. This was particularly the case with Buckland, for both in early and in later life he was always distinguished by his tact in illustrating extraordinary by common and familiar objects."

When, in 1814, Mr. Conybeare was about to leave Oxford for a country living, William Buckland, faithful to his scientific interests, hoped that the Suffolk parsonage " might prove to be founded on a bed of elephants."

Nor was it only at Axminster that Buckland, in his youthful days, found incentives to his pursuit of geological science. Speaking as President of the Geological Section of the British Association during its meeting at Bristol in 1836, he says that in the neighbourhood of Bristol he had learnt a part of his geological alphabet. " The rocks of this city were my geological school. They stared me in the face ; they wooed me, and caressed me, saying at every turn ' Pray, pray be a geologist ! ' "

At the age of thirteen the boy was sent to an ancient grammar school at Tiverton, founded in the seventeenth century by Blundell, a cloth manufacturer. A year later Mr. Pole Carew, Speaker of the House of Commons, obtained for him from Dr. Huntingford, the Warden, a nomination at Winchester. His uncle, the Rev. J. Buckland, Rector of Warborough in Oxfordshire, and Fellow of Corpus Christi College, Oxford, attracted by the ability of his nephew, advised his father to spare no expense in his education. " As William," he writes, " appears to excel your other boys by many degrees in talent and industry, he will probably make a better return for any extraordinary expense you may incur on this. occasion." To his uncle's judicious care and assistance Dr. Buckland doubtless owed much in his progress through life. A sagacious, energetic, stern-minded man, he was ever at his nephew's elbow, urging him to renewed efforts with encouragement, rebuke, and assistance.

As a boy at Winchester he became familiar, as he himself states, "with the chalk formation, from the fact of the pathway to the playground on St. Catherine's Hill passing close to large chalk pits, which abounded with sponges and other fossils, and from the practice of digging field mice from their holes in the surface of the chalk." Even in his schoolboy days he had already begun to collect objects of natural history, and was eager in the pursuit, or observation of the habits, of the mole-crickets, which abounded in the valley of the Itchen.

As a boy he was slow to learn, but what he once understood he never forgot. On one occasion, when he had regained several places which he had lost in class, the Head Master, Dr. Goddard, said to him, " Well, Buckland, it is as difficult to keep a good boy at the bottom of his class as it is to keep a cork under water." In later life he kept up his old Winchester associations by attending the yearly Wykehamist dinner in London, and he sent his sons—Frank, the well-known naturalist, and Edward, who was for many years in the Treasury—to that school. William Buckland's name may still be seen inscribed on a marble tablet upon the walls of the Seventh Chamber.

In 1801 he was elected Scholar of Corpus Christi

College, Oxford. He thus makes known his election in
a letter written to his father on May 13th : " I am happy
to inform you that I have just been elected the Senior
Scholar for Devonshire, after a course of many days'
rigorous examination against eight competitors." His
interests were already turning in the direction of geology.
" In my early residence at Oxford," he says himself, " I took
my first lesson in field geology in a walk to Shotover Hill
with Mr. Broderip, who knew much of fossil shells and
sponges from Mr. Townsend, the friend and fellow-labourer
of William Smith, ' the Father of English Geology.' The
fruits of my first walk with Mr. Broderip formed the
nucleus of my collection for my own cabinet, which in forty
years expanded into the large amount which I have placed
in the Oxford Geological Museum."

But although his special bent was, even in his undergraduate days, thus strongly developed, he did not neglect
the necessary studies of the University. In 1804 he took
his degree of B.A. He did not take honours, as there were
no class examinations in those days ; but he nevertheless
distinguished himself, for in a letter to his uncle he says :
" Before I came out of the schools they told me I had
passed extremely well, and after the Liceats were given
out they came up to me in the quadrangle, and said they
were extremely sorry they had not publicly thanked me
in the schools, but that I had passed a most creditable
examination."

His scholarship at Corpus, eked out by the income
derived from pupils, supported him for the next few years.
Meanwhile he was free to follow the course of studies in

which he was especially interested. "The interval," he writes, "between my Bachelor's and Master's degree afforded me leisure to attend the lectures of Dr. Kidd on Mineralogy and Chemistry, and of Sir Christopher Pegge on Anatomy; and my position as a Scholar of Corpus Christi College gave me the advantage of rooms and a small income from the College, which I augmented by taking pupils. Without the liberal aid of the endowments of the University, I could not have had the means which I enjoyed, during a residence of nearly forty-five years in Oxford, from April 1801 to December 1845, of acquiring knowledge during term time, and of enlarging it by extensive travelling during vacations."

In 1809 he was elected Fellow of his College, and in the same year was admitted into Holy Orders at the Chapel Royal, St. James's. Whether as a preacher or a tutor, Dr. Buckland, it may be mentioned, always wrote his sermons and lectures upon small slips of paper; and many years after, when preaching in the Chapel Royal, in the presence of the Queen Dowager, by some unfortunate accident the contents of his sermon case came fluttering down in all directions from the high, old-fashioned pulpit. The Doctor's old servant speedily came to the rescue, and, picking up the dispersed slips, handed them up to the preacher, who proceeded with his discourse, nothing disconcerted.

The vacations of his earlier Oxford time were often spent near Lyme Regis. For years afterwards local gossip preserved traditions of his adventures with that geological celebrity, Mary Ann Anning, in whose company he was to be seen wading up to his knees in search of fossils in the blue lias; "of his breakfast-table at his lodgings there, loaded with beefsteaks and belemnites, tea and terebratula, muffins and madrepores, toast and trilobites, every table and chair as well as the floor occupied with fossils whole and fragmentary, large and small, with rocks, earths, clays, and heaps of books and papers, his breakfast hour being the only time that the collectors could be sure of finding him at home, to bring their contributions and receive their pay; of his dropping his hat and handkerchief from the mail to stop the coach and secure a fossil; of the old woman who, finding him asleep on the top of the coach, relieved his pockets of a quantity of stones; of his travelling carriage, built extra strong for the heavy loads it had to carry, and fitted up on the forepart with a furnace and implements for assays and analysis."

Buckland's election to a fellowship enabled him to pursue those studies which made him, in the words of the historian and President of his College, "one of the most famous of English geologists, and indeed one of the creators of the science." His sitting-room, continues Dr. Fowler, was "a large room in the front quadrangle, now appropriated to the uses of an Undergraduates' Library, which was fitted up by him, irrespectively of personal comfort, as a Geological Museum—probably the earliest collection of the kind in Oxford, or perhaps in England, arranged on anything like scientific principles."

This is the room which, in its state of chaos, Mr. Philip Duncan so well describes in a poem dated May 1821 :—

"PICTURES OF THE COMFORTS OF A PROFESSOR'S ROOMS IN
C. C. C., OXFORD.

"Procul, este Profani,
Procul lucu."

"Away, ye ignorant and vain!
Away, ye faithless and profane!
Jesters and dainty dandies, fly hence!
But enter thou, dear son of science!
And here in mild disorder hurled
Behold an emblem of the world
In that chaotic state of old
When flints in Paramoudras rolled!
Here see the wrecks of beasts and fishes,
With broken saucers, cups, and dishes;
The præ-Adamic system jumbled,
With sub-lapsarian breccia tumbled,
And post-Noachian bears and flounders
With heads of crocodiles and founders;
Skins wanting bones, bones wanting skins,
And various blocks to break your shins:
No place is this for cutting capers
'Midst jumbled stones and books and papers,
Stuffed birds, portfolios, packing-cases,
And founders fallen upon their faces.
He'll see upon the only chair
The great Professor's frugal fare,
And over all behold illatum
Of dust and superficial stratum.
The sage amidst the chaos stands,
Contemplative, with laden hands,—
This grasping tight his bread-and-butter,
And that a flint, whilst he doth utter
Strange sentences that seem to say,
'I see it all as clear as day.'"[1]

---

[1] "Fugitive Poems connected with Natural History and Physical
Science," collected by the late C. G. Daubeny, M.D., F.R.S., F.G.S.
(Oxford: Parker & Son. 1861.)

LIFE OF DEAN BUCKLAND.

Another description of the same rooms—this time in prose—is given by Sir Roderick Murchison.

"On repairing," he says, "from the Star Inn to Buckland's domicile, I never can forget the scene that awaited me. Having, by direction of the janitor, climbed up a narrow staircase, I entered a long, corridor-like room, which was filled with rocks, shells, and bones in dire confusion, and in a sort of sanctum at the end was my friend in his black gown, looking like a necromancer, sitting on one rickety chair covered with some fossils, and clearing out a fossil bone from the matrix."

It is not, perhaps, surprising, especially if the social and intellectual conditions of the University at the beginning of the century be taken into consideration, that Buckland's conduct alarmed the older generation of College Fellows. Some dreaded lest his example should drive the *amœnitates academicæ* out of fashion; others suspected that the new studies might prove to be dangerous innovations. His goings and comings were therefore watched with an interest which was not wholly devoid of fear. When, in the early stages of his career, he started on a tour to the Alps and Italy—the results of which enabled him to produce one of the boldest and most effective of his writings—an authoritative elder is said to have exclaimed: "Well, Buckland is gone to Italy; so, thank God, we shall hear no more of *this geology.*"

The prophecy happily proved false, and Oxford dons were doomed to hear a good deal more of the obnoxious science.

CHAPTER II.

1808—1822.

" La vie des savans nous enseigne à chaque page que les grandes
vérités n'ont été découvertes et établies que par des études prolongées,
solitaires, dirigées constamment vers un objêt spécial, guidées sans cesse
par une logique méfiante et réservée."—CUVIER.

IN the summer of 1808 Buckland made his first geological tour. Alone and on horseback he travelled from
Oxford across the chalk hills of Berks and Wilts and
Dorset to Corfe Castle in the Isle of Purbeck. In the
vertical strata of hard white limestone on which that
castle stands he recognised the chalk, but the relations
of the strata above and below that formation were then
unknown. In the following year he explored in the same
way a large part of South Devon, visiting the granite of
Dartmoor, examining minutely the formations, and collecting specimens of the geology of the district. In 1810 he
made a tour through the centre and north of England,

11

examining the then unknown extent of the various strata, and colouring the results on Carey's large map of England. Other journeys followed in annual succession. Thus in 1813 Buckland, adopting the true Wykehamical fashion of going two and two, made a tour with his friend Mr. W. Conybeare in Ireland. In collaboration with his travelling companion, he published his first important paper—" On the Coasts of the North of Ireland." Among the organic remains in many of the chalk pits from Moira to Belfast and Larne he discovered some curious siliceous bodies known by the name of " Paramoudra." This word, which he could trace to no authentic source, Dr. Shuttleworth, late Warden of New College and Bishop of Chichester, contrived to hitch into verse if not into rhyme :—

> " When granite rose from out the trackless sea,
> And slate, for boys to scrawl—when boys should be—
> But earth, as yet, lay desolate and bare ;
> Man was not then,—but Paramoudras were."

No two of these curiously shaped pot-stones are exactly alike, their length commonly varying from one to two feet and their thickness from six to twelve inches. The substance of these bodies is in all cases flint ; in all cases, also, they have a central aperture, or pipe passing through their long diameter. They are found in different positions : sometimes they lie horizontally ; at other times they are inclined or erect. Buckland conjectured that the paramoudra may have possessed a character intermediate between a gigantic sponge [1] and an ascidian, and he

---

[1] Mr. W. Gray, M.A.F.A., of Mount Charles, Belfast, an authority

connected its mineral history with that of many other spongiform bodies which are found in the chalk flints. " Through the kindness," he writes, " of my learned friend Dr. Bruce, of Belfast, a very perfect specimen from Moira has been deposited in the Ashmolean Museum."

The origin of the word is, as has been said, uncertain. But the following story, told by the late Mr. Mant of Teddington, an Oxford pupil of Buckland's, gives an explanation of the term which is at least creditable to the quick-witted peasantry of Ireland.

On a hot, dusty day in Ireland, these " paramudras," as they should be correctly spelt, were first discovered as stepping-stones in a river. The Doctor apologised to his party that they must walk the rest of the journey, and that the stones must take up the carriage room. At the same time, taking a shilling out of his pocket, he asked a countryman what he called those stones. At first there was no response ; but when a second shilling appeared, Pat said, " Paramudras." When asked by his priest afterwards why he invented this word, " Faith," he replied, " the gentleman would have some name, and I hadn't one for him ; so para means ' against,' and paramudra (the stepping stone), ' against mud.' "

It was on this Irish tour that, after a very long and wet day among the cliffs, the two geologists, Conybeare and Buckland, entered at dark a lone hut, occupied by an aged female. Tired, hungry, and covered with mud and dirt,

upon flints, writes to the biographer (November 1893): " Dr. Buckland was the first to call them sponges, and the voyage of the *Challenger* has confirmed the correctness of the opinion."

they deposited their fossil bags and demanded refreshments. The old woman, much puzzled to make out their real character, set about her hospitable preparations. By the time they were complete, she had made up her mind. Placing the eggs and bacon on the table, she exclaimed : "Well, I never! fancy two real gentlemen picking up stones! What won't men do for money?"

Among his tours must also be mentioned the extensive journeys which he made with Mr. Greenough in the years 1812-15, for the purpose of collecting materials for the Geological Map of England. His letters frequently allude to this elaborate work, on which he was long engaged. Writing in April 1814 to Conybeare, he says :—

"I was not a little surprised to find from Greenough that he was in great hopes you would go with him to Paris to see Kings and Emperors, and Cuviers and Crocodiles. Should this actually take place, I need not, I trust, remind you to return loaded with a grand suite of specimens for the museum, and to establish a correspondence between Oxford and Paris, founded on an exchange of specimens. Illuminate Cuvier on the gypsum of Shotover, and press him to come and see us if he visits England. My lecture on the basin of Paris will be among the last of the set, so that you will be back in time to enrich it with your importation piping hot. I have made considerable progress with Serle in the last three days in arranging the specimens in the lower cabinets, from granite to mountain limestone. If you go to Paris, pray send me the notes you had begun touching Moses and Huttonianism, and which you took with you to finish, should there be opportunity. Send me also your map of Germany, if you do not take it with you, that I may transfer its contents to my map of Europe for the lectures."

In another letter, dated Corpus Christi College, April
29th, 1814, he writes to Conybeare :—

"The publication of Smith's map, which I sent for
yesterday, will preclude the necessity of my giving you
any trouble in finishing that which you had begun to
colour for me.   I have prepared and coloured sections
of the country round Oxford, and of the whole system
of the detail of the stratification of your grand section ;
for the last week I have been unpacking a barrel per
day, and have made considerable progress in the arrangement of the lower strata, assisted by Serle, who before
his departure last week disposed of the metals.

"I send you by the coach a parcel containing two maps
of mine and three of your own.   If you can possibly find
time, before Saturday in next week, to lay in the great
outlines of the mountain chains of Europe as you had
made them out in the map you took to town, I shall be
thankful if you forward it me by Saturday *the 7th* ; as
this is a matter of the first importance to me, you will, I
trust, have the goodness to take it in hand first.   With this
you will send me back the map of North America, having
written in the corner of it the explanation of the colours.
At your leisure you will much oblige me by inserting in
your map of France an outline of the chalk with its
superior formation, of the groups of granite and other
rocks which you know in that country.   I send with it your
map, which, from the lines inscribed on it, appears to
contain much of the requisite information, and add your
coloured map of the Netherlands, which will assist you in
the process.   Have you taken to town your sketches of
the coast at the Giants' Causeway, as they would be of
much service to illustrate the doctrine of subsidences ?
Pray send them me, if they are not in Oxford.   I suppose
you have had no leisure to think of Moses or Creation ? "

His enthusiasm was infectious.   Not only was he assisted
in these map surveys by his friends Conybeare, De la

Bêche, and Greenough; he had also enlisted the zealous interest of some ladies of high culture living at Penrice Castle, in the district of Glamorganshire known as Gower. In June 1815 he writes that "the information on the geology of Glamorganshire which he hoped to receive from the Lady Mary Cole and the Misses Talbot will be wanted for insertion in the Geological Society's Map of England, now far advanced in the hands of the engraver." He requests " that that part of the map of Glamorganshire which includes the hundred of Gower may be forwarded as soon as convenient," and hopes " to secure also the remaining parts of Glamorganshire, enriched by the geological observations of the approaching summer which he trusts the ladies will have the kindness to record upon it." He encloses " a drawing and description of some extraordinary coal-plants on the authority of an eye-witness, Mr. Walter Calverly Trevelyan, on whose father's property near Newbiggin they are found, and who promises to bring to Oxford in October drawings of every variety that he can find of fossil vegetables in that district." " These drawings will," he adds, " form a valuable subject of comparison with those of the South Wales coal-fields, should there be any collection of the latter in existence; if there be not, Mr. Buckland would venture to suggest to Miss Jane Talbot that she would afford an invaluable acquisition to the science of Botany and Geology, and acquire immortal reputation in both departments, by selecting a series of the most perfect fossil vegetables of the Welsh coal strata as her first essay in the noble art of lithography, for which also he hopes to bring her back some worthy

ANDREASBERG TO ELBINGRODE, SEPT. 17, 1822.

HALLE, 1822.

specimens from the mosses of the Carpathian Alps and Apennines."

In 1815 Buckland published the first comparative table

2

of the strata of England and those of the Continent, as arranged by Werner. This he enlarged in 1816, and distributed in Germany and France during a tour he made that year with John Conybeare and Greenough to Germany. This expedition was the first of a series of similar journeys,

PROF. BUCKLAND AND THE OCTOPUS.

in more than one of which Buckland was accompanied by Count Breüner. The Count was a skilful draughtsman, with a keen sense of humour, and it is to his pen that we owe the illustrations of episodes which occurred on a subsequent tour. In 1816 the travellers proceeded through Silesia to Poland, Austria, and Italy. From Weimar

Buckland writes : " We saw Goethe, and at Freyberg visited
Werner, who gave us a grand supper, and talked learnedly
of his books and music, and anything but Geology."

In another letter, written after his return to England,
he says : " The journey occupied five months of intense
labour, employed in seeing every collection and professor
that could be heard of, and purchasing every map, book,
and print that has been published relative to our favourite
science, or to the political economy of the countries we
passed through."

His friends at Penrice Castle were also kept informed
of his movements. In a long descriptive letter, written
in April 1817, Buckland tells Lady Mary Cole that he has
made " a rich collection of the shells of the Sub-Apennine
Hills, many of which resemble those of Hampshire and
Sheppey Island, and it would have been more perfect had
he not been arrested in the act of making it and sent back
fifteen miles to prison at Parma !" In spite of this
misfortune, he returned—

" highly satisfied with his tour, having accomplished every
point that was in contemplation before he set off.
Entering Hungary, he descended by the gold-mines of
Kremnitz and Schemnitz over a most picturesque country,
full of extinct volcanoes, to the great plain at the head of
which stands Presburg ; thence to Vienna, where are noble
collections in Natural History, by Styria and Carinthia
(countries equal to Switzerland in sublime Alpine scenery)
to Venice ; hence by the Euganean Hills (extinct volcanoes breaking up through chalk), Vicenza, Verona,
Mantua, and Parma, visiting by the way the fossil fish
quarries of Monte Bolca, which are in a formation above
and lying on chalk, and allied to the English Sheppey clay

and French *calcaire grossier.* Monte Bolca has also the same fossil plants as. Sheppey."

Another letter to Lady Mary Cole is too characteristic to be omitted.

" You have no doubt been wondering what is become of me and my projected tour into Glamorganshire, and I am sorry to inform you that all my movements have been deranged, and my plans thwarted, by an accident that befell me a month ago near Sidmouth, from the falling of an ignited spark of iron from my hammer into the

PROF. BUCKLAND VISITING MONTE BOLCA.

cornea of my eye, which I did not discover to be fixed there till some days after, when it began to oxydate. The result has been a series of five or six operations to cut out the minute rusty fragments, and a degree of inflammation which has prevented me from reading or writing during the last three weeks. I am happy to say the cause of injury is now totally removed, and in a few days I shall again take wing for Oxford. As I like always to extract all possible good out of the evil that befalls me, I have learnt two curious facts in physiology from my oculist at Exeter. First, that he once drew a tooth out of a patient's eye (literally an eye-tooth), growing between the bony

orbit and ball of the eye, and *I have seen the specimen.*
Second, that the belladonna leaf has the singular and
useful property, if laid on the eyelid, of causing a great
expansion of the pupil and iris, which is of the highest
service, in cutting for cataracts, to render visible the inner
chambers of the eye, and, in cases of diseased pupil, by
drawing the iris backwards in every direction, preserves it
from contact with the central injury.

"But, what is most important, I have been taught to
appreciate still more highly than I did before the value of
the organs of vision as the fairest inlets of knowledge and
pleasure to the soul.

"Passing yesterday over Kilmington Common on my
way to Exeter, I was at a loss to find a reason why a small
portion of that common is the only spot in England on
which the *Lobelia urens* has ever been found native.
Pray propose this as a hard question to Miss Jane, who
I know loves difficulties, and oblige me with her theory
on this subject. This is one of those curious questions
relating to the geography of plants for which I despair of
obtaining a satisfactory solution, unless from Humboldt or
herself. I fear I have imposed on her kindness a severer
task than I was aware of in asking for a nomenclature to
my marine plants, the value of which will be mightily
increased by her assistance in the arrangement of them."

Wherever he travelled, his eye was eagerly on the watch
for points at which he could observe geological strata, or
hunt for specimens. Quarries were irresistible attractions
to him, and, fortunately enough, he possessed a friend and
servant who tolerated his tastes. On the journey between
Oxford and Axminster, which he made once or twice
every year from 1812 to 1824, he rode a favourite old
black mare, frequently caparisoned with heavy bags of
fossils and ponderous hammers. She soon learnt her
duty, and seemed to take an interest in her master's

pursuits ; for she would remain quiet, without any one
to hold her, while he was examining sections and strata,
and then patiently submit to be loaded with the specimens
collected. Ultimately she became so accustomed to the
work that she invariably came to a full stop at a stone
quarry, and nothing would persuade her to proceed until
the rider had got off, and examined, or, if a stranger to
her, pretended to examine, the quarry. On one occasion
Dr. Buckland was in some danger from the falling stones
as he was climbing up the side of one of these quarries ;
when told of his danger by the bystanders—" Never
mind," said he, " the stones know me."

Buckland's enthusiastic labours were not without reward.
Not only the University of Oxford, but Lord Liverpool's
Government recognised the services of the man to whom,
in the words of Professor Brockhaus, " undoubtedly
belonged the honour of reducing the study of Geology to
a science."

In 1813 Buckland succeeded to the Readership of
Mineralogy which Dr. Kidd had resigned. The lectures
which the new Reader delivered in that capacity were
not confined to Mineralogy, but embraced the latest
discoveries and doctrines of Geology. His courses
attracted in a high degree the attention and admiration
of the University, and very largely contributed to the
public recognition of Geology as a science by the endowment in 1819 of a Professorship. The stipend of the
Professor was allotted from the Treasury, at the instance
of His Royal Highness the Prince Regent, for the delivery
of an annual course of lectures on Geology. To this new

office Buckland was appointed—a position which Sir Joseph Banks said "No one in England is so competent to fill." Writing in anticipation of this appointment to his friend Lady Mary Cole, in December 1818, he says :—

"DEAR LADY MARY,—I have just received a large importation from North America, and expect daily my stalactitic head and bones of Niobe from Venice. I hope soon to have a proper room prepared for their reception in Oxford, which is the least thing the University can do to meet the grant which you will be glad to hear I have obtained from the Crown for the establishment of a Professorship of Geology which I am to hold with my former office of Reader in Mineralogy. Nothing can exceed the strong exertions and flattering civilities I have received from Lord Grenville and Mr. Peel during the progress of this business, or the powerful representations which have been made to Lord Liverpool on my behalf. Sir Joseph Banks, too, on hearing of my Memorial to the Crown voluntarily requested Sir Everard Home to express to H.R.H. the Prince Regent that he felt great pleasure in the prospect of an establishment for Geology in Oxford, and considered no man in the country so proper to fill the situation as Mr. Buckland. Sir W. Scott and Lord Eldon have also given their assistance.

"I assure you I feel quite proud of the high consideration which is given to the noble subterranean science by such exalted personages, more especially by Lord Grenville, whom I am going to visit next week at Dropmore on my return to Oxford. During the last week I have been down to see Lord Tankerville's splendid collection at Walton, containing the finest shells and corals in the country, and extremely rich in fossil organic remains. He has a drawer full of Tortoises and Encrinites from the Sussex chalk, also of Pentacrinites from chalk, and lovely starfish. The plants in his hot-houses exceed in health and luxuriance any I have ever seen."

A letter written by Lord Grenville to Buckland in November 1820 speaks warmly of the latter's zeal in the cause of geology. " I am delighted," he writes, and the words are the more important as the writer was then Chancellor of the University of Oxford, "to learn that the interesting science which you are pursuing is making such rapid progress here and elsewhere, in consequence (I must say) in a very great degree of your indefatigable exertions."

To Buckland his " noble subterranean science," as he called it, appeared the most fascinating of pursuits, and his admiration for it was warmly expressed in the Inaugural Address which he delivered upon his appointment. In the course of his Address he acknowledges the "gracious encouragement" which His Royal Highness the Prince Regent had given to this infant establishment, and "the ardent zeal with which my application to the Crown on this occasion was furthered by Lord Grenville, the Chancellor of the University, whose care for good learning in this place it is impossible for me too highly to appreciate."

Modestly enough he pleads for Science forming a subordinate part in the University curriculum, while at the same time he would not surrender "a single particle of our system of classical study," which he regards as better than that prevailing on the Continent. " For some years past," he continued, "these newly created sciences have formed a leading subject of education in most universities on the Continent, and a competent knowledge of them is now possessed by the majority of intelligent persons in our country."

In reply to the arguments of utilitarians, who ask how far the science of geology can be made profitable, he observes :—

" The claims of geology may be made to rest on a much higher basis. The utility of Science is founded upon other and nobler views than those of mere pecuniary profit and tangible advantage. The human mind has an appetite for truth of every kind, physical as well as moral, and the real utility of science is to afford gratification to this appetite. The real question, then, more especially in this place, ought surely to be, how far the objects of Geology are of sufficient interest and importance to be worthy of this large and rational species of curiosity, and how far its investigations are calculated to call into action the higher powers of the mind. Now when it is recollected that the field of the geologist's inquiry is the Globe itself; that it is its study to decipher the movements of the mighty revolutions and convulsions it has suffered, convulsions of which the most terrible catastrophes presented by the actual state of things (earthquakes, tempests, volcanoes), afford only a faint image,—the last expiring efforts of those mighty disturbing forces which once operated,—these surely will be admitted to be objects of sufficient magnitude and grandeur to create an adequate interest to engage in their investigations."

With arguments forcibly and clearly stated, the Professor goes on to show how Geology, which he regarded as the handmaid of Religion, holds the keys of one of the kingdoms of Nature ; how closely it is allied to Mineralogy and Chemistry ; how it can claim the application of pure Mathematics ; and how it is connected with Hydrostatics and with Astronomical speculations. And then, passing into a higher region, he points out its connection with Natural Theology, and shows that the working of the Great

First Cause is not less demonstrable from the structure of the earth than are His wisdom, power, and goodness.

This lecture was afterwards published under the title of "Vindiciæ Geologicæ ; or, The Connection of Geology with Religion explained." The object of his lecture was to show that the study of Geology, so far from being irreligious or atheistic in its consequences, had a tendency to confirm the evidences of Natural Religion ; that there could be no opposition between the works and the word of God ; and that the facts developed by it were consistent with the accounts of the Creation and the Deluge as recorded in the Book of Genesis. The inaugural lecture may still be read with pleasure for the ability and elevated feeling with which the Professor defended Geology, and every other science, from the narrowness of utilitarians. But while arousing interest he also excited opposition, and every onward step that he made towards giving the science of Geology a position in the University created opponents to its claims. Sometimes the opposition was serious enough, his opponents being men who feared that the study of God's earth would shake the foundations of Christianity ; sometimes the objections raised only elicited a hearty laugh from the Professor. His friends had their jokes at the expense of the enthusiastic geologist. Here, for example, is a couplet suggested by Pope's on Sir Isaac Newton, from the pen of Shuttleworth :—

> "Some doubts were once expressed about the Flood ;
> Buckland arose, and all was clear as—mud."

Deeply engrossed though Professor Buckland was in geological pursuits, they were far from exclusively absorbing his interests. Every project for improving or advancing the condition of the University and the City of Oxford as a place of residence received his careful attention. In 1818, in the face of strong opposition, he succeeded, with the aid of several influential men, in lighting Oxford with gas, and was for many years chairman of the Company. He also did good service to the city by promoting plans for the improvement of the sewerage and of the water supply. His labours as a sanitary reformer were indeed unremitting, and the experience thus gained by the Professor at Oxford was, as will be seen later on, turned to excellent account by him as Dean of Westminster.

It is always the busiest men who know how to find leisure, and Buckland's advice and active assistance were asked for a variety of good deeds, and never asked in vain. His power of work and his willingness to help were indeed well-nigh inexhaustible. If his ardour sometimes made him a little impatient, his genuine kindness of heart, combined with a keen sense of humour, speedily corrected the momentary impulse. However strong his convictions, he was never so wedded to his own judgment as to shrink from opposition.

As Professor his classes at Oxford were always well attended, and his genial good-humour and apt description of things around him made every one happy, and therefore in a humour to listen, learn, and recollect.

Outside the University his gifts as a lecturer were also warmly appreciated. Miss C. Fox records in her journals that Buckland says " he feels very nervous in addressing

large assemblies till he has once made them laugh, and
then he is entirely at his ease." He always liked to have
a picture to show his audience, where specimens were
not available, and in a letter to Sir Henry de la Bêche,
he says : "With respect to a block (engraving), it was
ready to go to the printer to-morrow if you approved ;
though bad, it is better than nothing, and I like always
to tell my story by a picture if possible."

Buckland, whilst staying with Miss Fox, one wet day
gave a lecture in the drawing-room. "We listened," Miss
Fox says, "with great and gaping interest to a description
of his geological map, the frontispiece to his forthcoming
Bridgewater Treatise. He gave very clear details of the
gradual formation of our earth, which he is thoroughly
convinced took its rise ages before the Mosaic record. He
says that Luther must have taken a similar view, as in the
translation of the Bible he puts ' 1st ' at the third verse of
the first chapter of Genesis, which showed his belief that
the two first verses relate to something anterior. He explains the hills with valleys between them by eruptions
underground. He compared the world to an apple-dumpling,
the fiery froth of which fills the interior, and we have just
a crust to stand upon ; the hot stuff in the centre often
generates gas, and its necessary explosions are called on
earth volcanoes. He gave descriptions of antediluvian
animals, plants, and skulls. They have even discovered a
large fossil-fish with its food only partially digested."

A characteristic story is related of Buckland, to the effect
that he and a friend, riding towards London on a very dark
night, lost their way. Buckland therefore dismounted,

EXPEDITION TO SHOTOVER.

and taking up a handful of earth, smelt it. "Uxbridge," he exclaimed, his geological nose telling him the precise locality. He was very fond of " field lectures " as an adjunct to his ordinary course, and they were always well attended, both by students and others interested in the practical study of geology. On one occasion, when lecturing on Shotover Hill, a member of his class, Mr. Howley, afterwards Archbishop of Canterbury, discovered a lark's nest with eggs in it, and bringing it to the lecturer, asked if he considered it to be of the " oolite formation." Buckland also delighted in giving a new class of equestrian listeners a practical lesson in geology, by sticking them all in the mud to make them remember the Kimmeridge clay. He would often give out as a notice at the end of a lecture, " To-morrow the class will meet at the top of Shotover Hill at ten o'clock "; or, " The next lecture will take place in the fields above the quarry at Stonesfield "; or, " The class will meet at the G. W. R. Station at nine o'clock ; when, in the train between Oxford and Bristol, I shall be able to point out and explain the several different formations we shall cross ; and, if you please, we will examine the rocks and some of the most interesting geological features of Clifton and its neighbourhood." The true meaning of the terms " stratification, denudation, faults, elevations, etc., could never be learnt in a lecture-room," he would say.

Sir Henry Acland, one of the few of Buckland's pupils still living, tells a characteristic story of his manner of lecturing. It shall be given in his own words :—

" You have asked me," writes Sir Henry to the biographer, " to tell you how I was attacked by Professor Buckland,

and why, in the middle of one of his exciting and graphic lectures.

" It was in this wise—and, as you desire it, the whole story must be told.

" When I was a boy my father took me to Sir Benjamin Brodie and said, ' Sir Benjamin, I want to make this boy a physician. What is to be done?' I was frightened out of my wits as the eagle-eyed man looked at me from head to foot. He replied, ' What is your University?' ' Oxford,' said my father. ' Send the boy to Oxford,' said the great surgeon quickly. ' While there he is not to attend to anything connected with his future profession, but be as though he was to be like you in Parliament. When he has taken his degree, let him come to me, and I will tell him what to do then.' In another minute we were out of the room. Some fifty were waiting elsewhere.

" When in 1835 I went to Christ Church, your father got hold of me, being very friendly with mine ; assured me geology had nothing to do with medicine, and bade me attend his lectures.

" I can never forget my *début* as his pupil—though it was not our first acquaintance, for I had made diagrams for his great evening address at the British Association in Edinburgh in 1834, and knew his ways.

" He lectured on the Cavern of Torquay, the now famous Kent's Cavern. He paced like a Franciscan Preacher up and down behind a long show-case, up two steps, in a room in the old Clarendon. He had in his hand a huge hyena's skull. He suddenly dashed down the steps— rushed, skull in hand, at the first undergraduate on the front bench—and shouted, ' What rules the world?' The youth, terrified, threw himself against the next back seat, and answered not a word. He rushed then on me, pointing the hyena full in my face—' What rules the world?' ' Haven't an idea,' I said. ' The stomach, sir,' he cried (again mounting his rostrum), ' rules the world. The great ones eat the less, and the less the lesser still.' "

The Professor's *forte* as a lecturer in these early days

excited the rhyming propensities of his College friend
Shuttleworth, afterwards Warden of New College, and
subsequently Bishop of Chichester. The lecture which
suggested the following lines was probably delivered
early in 1822.

> "In Ashmole's ample dome, with look sedate,
> 'Midst heads of mammoths, heads of houses sate;
> And tutors, close with undergraduates jammed,
> Released from cramming, waiting to be crammed.
> Above, around, in order due displayed,
> The garniture of former worlds was laid:
> Sponges and shells in lias moulds immersed,
> From Deluge fiftieth, back to Deluge first;
> And wedged by boys in artificial stones,
> Huge bones of horses, now called mammoths' bones;
> Lichens and ferns which schistose beds enwrap,
> And—understood by most professors—trap.
> Before the rest, in contemplative mood,
> With sidelong glance, the inventive Master stood,
> And numbering o'er his class with still delight,
> Longed to possess them cased in stalactite.-
> Then thus with smile suppressed: 'In days of yore
> One dreary face Earth's infant planet bore;
> Nor land was there, nor ocean's lucid flood,
> But, mixed of both, one dark abyss of mud;
> Till each repelled, repelling by degrees,
> This shrunk to rock, that filtered into seas;
> Then slow upheaved by subterranean fires,
> Earth's ponderous crystals shot their prismy spires;
> Then granite rose from out the trackless sea,
> And slate, for boys to scrawl—when boys should be—
> But earth, as yet, lay desolate and bare;
> Man was not then,—but Paramoudras [1] were.
> 'Twas silence all, and solitude; the sun,
> If sun there were, yet rose and set to none,

---

[1] Paramoudras: see page 13.

LECTURE IN ASHMOLEAN, 1822.

[Face p. 32.

Till fiercer grown the elemental strife,
Astonished tadpoles wriggled into life;
Young encrini their quivering tendrils spread,
And tails of lizards felt the sprouting head.
(The specimen I hand about is rare,
And very brittle; bless me, sir, take care!)
And high upraised from ocean's inmost caves,
Protruded corals broke the indignant waves.
These tribes extinct, a nobler race succeeds:
Now sea-fowl scream amid the plashing reeds;
Now mammoths range, where yet in silence deep
Unborn Ohio's hoarded waters sleep.
Now ponderous whales . . .
        (Here, by the way, a tale
I'll tell of something, very like a whale.
An odd experiment of late I tried,
Placing a snake and hedgehog side by side;
Awhile the snake his neighbour tried t' assail,
When the sly hedgehog caught him by the tail,
And gravely munched him upwards joint by joint,—
The story's somewhat shocking, but in point.)
Now to proceed:—
The earth, what is it? Mark its scanty bound,—
'Tis but a larger football's narrow round;
Its mightiest tracts of ocean—what are these?
At best but breakfast tea-cups, full of seas.
O'er these a thousand deluges have burst,
And quasi-deluges have done their worst.'"

The lecture ends with a couplet which the facetious writer' observes of its own accord "slides into verse, and hitches in a rhyme" :—

> " Of this enough. On Secondary Rock,
> To-morrow, gentlemen, at two o'clock."

Another witness to Buckland's impressiveness as a lecturer was Colonel Portlock, President of the Geological Society in 1875.

" His invariable cheerfulness," he writes, " and humour threw light over the description of any subject he took in hand ; and whether describing with his pen or with his tongue, the ancient inhabitants of the earth, such was the vivid reality of the picture that he drew, that they appeared to act and speak before us, so that we may fairly designate him the Æsop of extinct animals—alas ! himself now extinct ! how can we hope to see again in all its fulness a second Buckland ?　To form a correct notion of the powerful manner in which Dr. Buckland influenced the progress of Geological Science, it would be necessary, not only to pass in review the long series of his geological contributions, but also to realise the effect he produced on his hearers, and on the University generally, by his lectures. It is impossible to convey to the mind of any one who had never heard Dr. Buckland speak, the inimitable effect of that union of the most playful fancy with the most profound reflections which so eminently characterised his scientific oratory.　To him more than to any geologist are we indebted for unexpected suggestions, curious inquiries, and novel kinds of evidences."

Frank Buckland writes, in his account of the sale in January 1857 of his father's minerals, fossils, etc. : " There was great competition for the hammers ; these relics are much prized by the possessors, for by means of them my father hammered out much information from the breast of mother earth."　Mr. Etheridge tells the story of Buckland when travelling in Scotland, in order not to shock the feelings of the Scotchmen on Sunday, carrying his hammer up his sleeve.

The charm that marked Buckland's lectures was felt also in his character and conversation.　When Mr. Ruskin was an undergraduate of Christ Church, the Professor of Geology was a Canon of the Cathedral.

"There was," says Mr. Ruskin in his "Præterita,"[1] "a
more humane and living spirit, however, inhabitant of the
N.W. angle of the Cardinal's square ; and a great many of
the mischances which were only harmful to me through
my own folly may be justly held, and to the full, counter-
balanced by that one piece of good fortune, of which
I had the wit to take advantage. Dr. Buckland was a
Canon of the Cathedral, and he, with his wife and
family, were all sensible and good-natured, with originality
enough in the sense of them to give sap and savour to
the whole College. . . . All were frank, kind, and clever,
vital in the highest degree ; to me, medicinal and saving.
*Dr. Buckland was extremely like Sydney Smith in his staple
of character ; no rival with him in wit, but like him in
humour, common sense, and benevolently cheerful doctrine of
Divinity.* . . . Geology was only the pleasant occupation
of his own merry life."

Another distinguished Oxonian speaks enthusiastically
of Buckland's vivacity, mirthfulness, and power as a
talker. Writing in 1892, Professor Storey Maskelyne
says :—

"Dr. Buckland's wonderful conversational powers were
as incommunicable as the bouquet of a bottle of champagne,
but no one who remembers them as I do, can ever forget
them.
"It was indeed at the feast of reason and the flow of
social and intellectual intercourse that Buckland shone.
'A merrier man within the limit of becoming mirth I
never spent an hour's talk withal.' Nothing came amiss
to him, from the creation of the world to the latest news
in Town ; from the flora and fauna of ages long past to
the last horticultural meeting at Chiswick or Exhibition
at the Zoological Gardens ; through all intermediate time
he was equally at home. *Nihil tetigit quod non ornavit,*

---

[1] Ch. xi., pp. 375, 376-7, 381.

there were few subjects which he could not more or less illustrate. In build, look, and manner he was a thorough English gentleman, and was appreciated in every circle."

Sowerby has recorded an anecdote of Buckland galloping off with a huge ammonite over his shoulders, his head passed through the opening occasioned by the loss of the central volution, when his companions dubbed him on the spot "Ammon Knight." "A man of devout spirit, strong of mind and strong in body, working hard and setting others to work, gathering and giving knowledge, a patient student, a powerful teacher, a friendly associate, a valiant soldier for geology in days when she was weak, an honoured leader in her hour of triumph."

One of the most notable and lasting of scientific friendships was formed between two of his pupils, Sir Philip de Malpas Grey Egerton, Bart., and Viscount Cole (afterwards Lord Enniskillen), both of whom were at Christ Church at that time. In the Long Vacation of 1820 these two young men set out on their geological travels through Europe. Dr. Buckland sent them first—after William of Wykeham's fashion of "two and two"—to collect bones and work out for him the latest discovered cave in Bavaria. Mr. Etheridge, of the Natural History Museum, says that, before starting on their journey, these two friends made their wills. In case of the death of either, the joint collection was to belong to the survivor for his life ; on his death the collection was to be sold, and offered first to the British Museum, then to their Alma Mater of Oxford, after that to Cambridge and Paris. If not purchased by any of these bodies, the world in general was to have the option of buying. The Americans would have given twice the sum for these valuable and unique specimens; but they were purchased by the British Museum for a sum of several thousand pounds. Sir Philip Egerton's brother, the Rev. W. H. Egerton, Rector of Whitchurch, Salop, writes that " the bulk of both collections consisted of fossil fishes. When a slab containing a specimen was split in halves the two friends tossed up for first choice, the one half containing the bones of the fish—the other the impression. This was the case with a vast number of specimens chiefly from Solenhofen—the two collections being brought together at Kensington form a complete whole."

The ample vacations which Buckland enjoyed as an Oxford Professor enabled him to continue his geological tours at home and abroad. Thus in 1820 he made an expedition to France with his friend Conybeare. Writing from Lyons to Sir John Nicholl, he says :—

" Three days brought me from London to Paris, where my first business was to call on Cuvier, who after receiving me with the greatest cordiality, and saluting my cheeks with more than English familiarity, immediately made a dinner for me, inviting Humboldt, Biot, Cordier, Bowditch the African traveller, Frederick Cuvier, and several others of the savants of Paris, and giving me admission to the entire establishment of the Jardin du Roy. I attended three lectures on geology by Cordier, two on entomology by La Traille, and three on ornithology by Geoffrey St. Hilaire. I admired exceedingly the French style of lecturing ; the manner and matter were extremely good, but the classes as ill-looking and ungentlemanly a set of dirty vagabonds as ever I set eyes on, and not more numerous than my own at Oxford. I attended also a meeting of the Institute

at which was announced the death of poor Sir Joseph
Banks, who is not less regretted in France than in our own
country. I saw there Guy Lusac, Menard, Vaguelin,
Henry Raymond, Brockard, Bindon, and most of the first
scientific men of France, whose love of Science, however, does not induce them to attend without receiving
about eight shillings a head for their hour's work.

"I find the best geologists in France to be Cordier, the
successor of Fangas St. Ford as lecturer in geology, and
Bindon, who is curator of the King's collection under
Count Bourdon, and is on the point of publishing an
excellent work on the geology of Hungary, with a map
and lectures, that will be extremely good, for he thoroughly
understands his work. He was sent to Hungary by the
King two years ago. I find them all most deplorably
deficient in knowledge of their country, as well as in
general geology. Our Society would number at least
thirty members that would beat the best of them, and
never did I feel myself more highly gratified in the article
of pride than I was by the manner in which they flocked
round me to propose their difficulties, and the passive
obedience with which they received my oracular decisions.

"I saw a great deal of Humboldt, whom I liked exceedingly, and with whom I am likely from henceforth to be
in continual correspondence. He talks more rapidly and
more sensibly than any man I ever saw, and with a
brilliancy that is indicative of the highest degree of genius.
He is on the point of publishing a most interesting work,
a comparative view of the geological structure of Europe
and South America, and, according to the documents he
showed me, the identity of the phenomena of the two
continents is more absolute than the most sanguine wishes
could have anticipated. He has given me a section of
the valley of Santa Fè de Bogota, which is the exact
counterpart of the valley of Glamorganshire, which I shall
publish with my account of the Severn district in our
Transactions. He will make use of my list of the order
of succession of English strata, and in almost all points
but the history of the Old and New Red Sandstone, which

Taking leave of Sir J[oseph] & Lady Banks

Underwood Geologerum Primigus

Robert Brown   Sir Joseph Banks   W.D Conybeare   W. Buckland

is the great stumbling-block of continental geologists, we
are fully agreed. On this, however, I have made a convert
of Bindon, and hope soon to convince Humboldt.

"I left Paris with most pressing invitations to visit it
again on my return, having allowed myself time to attend
to nothing there but my undergroundology, and dashed
directly into Auvergne. At Clermont I made a considerable collection of petrified fruit baskets, and took the
tour of the volcanic chain and summit of Puy-de-Dôme.
It is the finest thing by far in Europe; according to
Humboldt it exactly resembles a similar chain in Mexico,
and presents more than fifty craters nearly in a line from
north to south, many of which are larger and finer than
that of Vesuvius. The streams of lava also are not less
decided ; one of them is three miles broad and six miles
long. They are all post-diluvian, though there are no
records of the time when they were in action ; they stand
on, and have burst up through, an enormous mass and
elevated plain of granite, which is covered first by trap
and this again by lava. The portion of Clermont is,
perhaps, the finest thing in France, and the mountains I
have crossed between Clermont and Lyons, being entirely
granitic, are yet beautiful, presenting that second-rate style
of mountain scenery which we have in the best part of
Monmouthshire. I am disappointed in Lyons, because
I had heard too much of it. It is certainly a bad thing
to have too good a character."

In this letter he alludes to the death of Sir Joseph
Banks. Before starting on his journey he had called to
take leave of this famous patron and encourager of travellers and science. He never saw him again alive, and it
is this farewell interview which Count Breüner has cleverly
sketched. Sir Joseph was then much invalided with the
gout, but, though a martyr to the complaint, he is said
to have had such self-control that he never showed any
irritability. Both at Christ Church and at Islip Buckland

planted yellow banksia roses in memory of his friend, who had been for forty-one years President of the Royal Society.

Buckland was greatly pleased, on his return from this long sojourn on the Continent, to be greeted with the following epitaph written by his friend Whately, afterwards the famous Archbishop of Dublin. He had the verses lithographed, and gave copies to his friends, so that they are more known than many of the clever verses written by Dr. Shuttleworth and Mr. Duncan.

ELEGY

Intended for Professor Buckland. December 1st, 1820.

BY RICHARD WHATELY.

" Mourn, Ammonites, mourn o'er his funeral urn,
  Whose neck ye must grace no more ;
Gneiss, granite, and slate, he settled your date,
  And his ye must now deplore.
Weep, caverns, weep with unfiltering drip,
  Your recesses he'll cease to explore ;
For mineral veins and organic remains
  No stratum again will he bore.

"Oh, his wit shone like crystal; his knowledge profound
  From gravel to granite descended,
No trap could deceive him, no slip could confound,
  Nor specimen, true or pretended ;
He knew the birth-rock of each pebble so round,
  And how far its tour had extended.

" His eloquence rolled like the Deluge retiring,
  Where mastodon carcases floated ;
To a subject obscure he gave charms so inspiring,
  Young and old on geology doated.
He stood out like an Outlier; his hearers, admiring,
  In pencil each anecdote noted.

"Where shall we our great Professor inter,
     That in peace may rest his bones ?
If we hew him a rocky sepulchre,
     He'll rise and break the stones,
And examine each stratum that lies around,
For he's quite in his element underground.

"If with mattock and spade his body we lay
     In the common alluvial soil,
He'll start up and snatch those tools away
     Of his own geological toil ;
In a stratum so young the Professor disdains
That embedded should lie his organic remains.

"Then exposed to the drip of some case-hardening spring
     His carcase let stalactite cover,
And to Oxford the petrified sage let us bring
     When he is encrusted all over;
There, 'mid mammoths and crocodiles, high on a shelf,
Let him stand as a monument raised to himself."

Almost at the same time when Dr. Buckland was
making these extensive tours in Great Britain to collect
materials for a geological map of England, and in foreign
countries to procure valuable and unique specimens for
his museum, a kindred spirit was inaugurating a similar
movement in America. A young Scotch merchant,
William Maclure, born in Ayr, author of the "Pioneers of
Discovery," went forth, with his hammer in his hand and
his wallet on his shoulder, to make a geological survey
of the United States. Pursuing his researches in every
direction, often amid pathless tracts and dreary solitudes,
he crossed and recrossed the Alleghany Mountains no less
than fifty times. He encountered all the privations of
hunger, thirst, fatigue, and exposure, month after month,
year after year, until his indomitable spirit had conquered
every difficulty and crowned his enterprise with success.

It was during his journeys, at home and abroad, that
Buckland laid the foundation of a collection which became
famous as the first of its kind in Europe. In its formation he expended a large portion of his private fortune,
and if there was a good specimen to be anywhere
obtained, he would secure it at any price. The collection
of cave bones from England and the Continent is unique.[1]
The other specimens were selected with a view to their
fitness for illustration of certain definite points. Some
are of the most delicate texture ; others again are of
such gigantic size and ponderous weight that they show,
as Professor Phillips remarked, " the courage of the man."

Not only did he spend his own money, time, and strength
in the formation of his collection ; friends were also
working for him in all parts of the globe. Writing in
1819 to Lady Mary Cole, he says :—

" My treasures in Geology continue more than ever to
accumulate. I have just heard from town that three large
Russian boxes from Mr. Strangways are sent off to Oxford,
and in his last package I received a diploma from Moscow,
for which I am indebted to his kindness. You will be
pleased to hear I am likely to get extensive importations
from all the British colonies over the world, through the
kindness of Lord Bathurst, who lately sent me a message
requesting I would draw up a list of instructions for
collecting specimens in Geology, of which he would
transmit copies to all the colonies connected with his
office, and adding that it is his intention to deposit the
specimens that may be sent home for the purpose of
illustrating my lectures."

---

[1] F. Buckland's memoir of his father, prefixed to the "Bridgewater
Treatise," 3rd edition.

And the collection, adds Buckland, "is becoming one of the most valuable in Europe."

Among the specimens with which his friends enriched his collection were treasures gathered from the Arctic regions by adventurous explorers. In the progress and results of the various Polar expeditions he was keenly interested. With most of the officers who were engaged he was personally acquainted; to more than one he had given valuable assistance in the preparation of geological reports; and in the classification and arrangement of their collections his aid was often invoked. It was therefore an appropriate tribute to his geological services when his name was bestowed, by Captain Beechey, on a new-found island and a newly discovered river.

Of Captain Ross's expedition he writes, on December 14th, 1818, a long account to Lady Mary Cole at Penrice Castle :—

"The philosophical world here," he says, "is much occupied with the question of the Polar expedition; Captain Ross and his under officers give different accounts, and three books are in preparation. I saw Captain Ross a few days ago at Sir Joseph Banks', and was at the British Museum on the arrival of the animals and boxes of specimens. Captain Ross had a chart of Baffin's Bay corrected by daily soundings and observations. At the extreme point which they reached, after sounding in calm water 1,000 fathoms, the sea shallowed gradually to 300, and a lofty ridge of mountains on the right, as they sailed forward, seemed to close round and shut up the end of the bay. Of this he had little doubt; but his officers thought otherwise, and they sailed in at evening, hoping to establish the fact. But at night a gale came on, and they were obliged to turn back, and were never again able to enter this bay, I believe, from ice or fogs which followed the storm. So there remains still a point at which land has not been seen, and a possibility of a passage, but no probability.

"On turning the ship to come out of this bay, the needle, which had been steady as they sailed inwards before the wind, became exceedingly irregular as soon as they began to beat against it. Something of this they attribute to the form of the ships, and two ships, differently constructed from each other, are to repeat next year the experiments that have been made. The question is still undecided whether Greenland be an island, and it highly becomes this country to ascertain the point, if possible, and correct the charts of the Polar seas.

"Near the north end of Baffin's Bay, on the east side, at Lake Sir Dudley Digges, along six miles of coast they found extensive irregular patches of red snow on the country of the new tribe of Esquimaux, whose language was much more intelligible to the interpreter who accompanied the expedition than he was to the ship's crew. There seems little doubt of the colouring matter being caused by birds, which swarm on this coast in one small pool of water amid an ocean of icebergs. Captain Ross told me his boat's crew shot, in four hours, 1,600 birds, which were drawn to this as the only spot where they could find their food, consisting of shrimps and medusa, which also constitute the food of the whales. Fish are rare in these cold latitudes. The birds are beautiful —chiefly puffins, gulls, auks, guillemots, of which great numbers are imported ; many specimens of the ivory gulls, which are extremely rare ; also marine animals of the lower orders.

"Red snow may be seen in rabbit warrens, and De Saussure mentions it in the Alps at Mt. Breven and St. Bernard. The Bishop of Oxford and Mr. Honey have also seen it on the Alps. De Saussure wishes to believe it the pollen of plants, but is at a loss where to find the plants. He says it only stains the surface to the depth of a few inches, not exceeding three, and seems to be a fine

powder washed down into the hollows of the snow. It
has been suggested that the colouring matter may be
derived from the lichen Tartareus (Roccella or Orchill of
commerce), which is imported largely from Corsica and
Sardinia for the use of dyers ; but as this plant must be
steeped in a solution of ammonia to extract its red colour,
and is not likely to meet that substance on the high Alps,
we must refer the colour to the same source as in Baffin's
Bay. I believe the colour obtained from this lichen is
called Cudbear, and enclose a specimen of it for Miss Jane,
which comes from Scotland ; it is most luxuriant on granite
rocks. I must request her indulgence for all the errors
I may have made touching the history of this lichen, and
shall hope to be corrected where I am wrong. My friend
Mr. Duncan has another theory more pretty than any of
the rest, if it were but true, and which he has committed to
verse as follows :—

> "'Of yore 'tis said a heavenly red
> The cheeks of modest maids o'erspread :
> Some say with innocence it fled—
> But where it went no man could know ;
> The truth our modern travellers show—
> It went to dye the Arctic snow.'

" The natives of this poetically coloured region, says Sir
Dudley Digges, have harpoons and knives made of iron
beat flat between two stones. This iron no doubt fell
from the clouds, like the mass of native iron found by
Pallas in Siberia ; and it is only in such cases that
malleable iron has been found native, being always
accompanied by nickel. That used in the harpoons has
three-hundreds of nickel ; a small knife has been made in
London from twenty-six grains of it. Captain Ross could
not make out the size of the block from which the natives
obtained it, nor its position ; but it is beyond doubt
meteoric.[1]

" Blocks of fir and fragments of ships are drifted occasion-

---

[1] Baron Nordenskiold has proved that this is native and not meteoric
iron.

ally on the coast; but the natives find the bone of whales and teeth of walrus best calculated, from strength and lightness, to make their sledges, of which a good specimen is brought home. They burn whale oil, making a wick of moss, which serves also for fuel. They have scarcely any plants but mosses, and no quadrupeds except bears, hares, dogs, and white foxes. The dogs resemble wolves with short legs, are very strong, well fitted for sledge harness, and perfectly gentle. There is a live fox which I saw this morning at the Museum basking in the hoar frost; he prefers staying on the outside of his house in the coldest nights, and is quite white and by no means savage.

"As to the rocks, the west coast of Baffin's Bay, on the only spot they touched, resembled Derbyshire in its limestone and trap. The blocks floating on the icebergs were chiefly granite mica, slate, and trap; and the coast of Greenland near Disco, trap with a bed of imperfect coal in it.

"The other ships have not done so much as those from Baffin's Bay. I have seen none of their officers, but have been on board the *Alexander*, which was with Captain Ross, and obtained specimens of the mosses, which I will soon forward to Penrice. The Spitzbergen ships were impeded by their accident from proceeding : one of them was right between two masses of ice, and raised out of the water; her side was forced in, and a barrel of meal within pressed flat as a pancake. The only mode of repairing her was lashing her to an iceberg, and pulling her mast downwards, until her side rose out of the water sufficiently to have planks laid on the outside. She was much too damaged to proceed. I saw yesterday at Murray's some drawings which will be engraved of the situation of the ships in a storm amidst the icebergs, dashing every minute against enormous floating rocks of ice, from which it seems miraculous how they ever could have escaped."

In May 1825 the *Blossom*, under Captain Beechey, was sent out to afford such assistance as might be required by

Captains Franklin and Parry, who had started the previous
year on a second voyage of discovery to the Arctic regions.
During the autumns of 1826 and 1827, Captain Beechey
was to await in Behring Straits the appearance of one
or both of these officers. As his vessel would have to
traverse in her route a portion of the globe hitherto little
explored, it was intended to employ her in surveying and
exploring such parts of the Pacific as were within her
reach, and for this purpose the ship was provided with
both naturalist and surveyor. In the group of the Bonin
Islands, Captain Beechey found one composed of basaltic
pillars ; " far grander," he writes to Buckland, " than the
Giants' Causeway." He named it Buckland Island and
adds that " on the south side it is possessed of a good
harbour." In July 1826 the *Blossom* anchored in Kotzebue
Sound, there to await the arrival of Captain Franklin.
Captain Beechey employed the time in surveying and
exploring as much of the coast as possible. He visited
the extraordinary ice formation in Eschscholtz Bay mentioned by Kotzebue as being " covered with a soil half
a foot thick, producing the most luxuriant grass," and
" containing an abundance of mammoth bones." Sailing
up the bay, which was extremely shallow, he landed at a
deserted village on a low sandy point, to which he gave
the name of Elephant Point, from the bones of that animal
being found near it.

" The cliffs in which this singular formation was discovered begin near this point, and extend westward in
a nearly straight line to a rocky cliff of primitive formation
at the entrance of the bay. The cliffs are from twenty to

eighty feet in height, and rise inland to a rounded range of hills between four and five hundred feet above the sea."

Leaving Mr. Collie, the ship's surgeon, with a party to examine the cliffs in which the fossils and ice formation had been seen by Kotzebue, Captain Beechey proceeded to the head of the bay in a smaller boat.

"We landed upon a muddy beach, and were obliged to wade a quarter of a mile before we could reach a cliff for the purpose of having a view of the surrounding country. Having gained its summit, we were gratified by the discovery of a large river coming from the southward and passing between our station and a range of hills. At a few miles' distance the river passed between rocky cliffs, whence the land on either side became hilly, and interrupted our further view of its course. The width of the river was about a mile and a half, but this space was broken into narrow and intricate channels by banks, some dry and others partly so; the stream passed rapidly between them, and at an earlier period of the season a considerable body of water must be poured into the sound, though from the comparative width of the channels the current of the latter is not much felt. The shore around us was flat, broken by several lakes, in which there were a great many wild-fowl. The cliff we had ascended was composed of a bluish mud and clay, and was full of deep chasms.

"Meanwhile Mr. Collie had been successful in his search among the cliffs at Elephant Point, and had discovered several bones and grinders of elephants and other animals in a fossil state. Associating these two discoveries, I bestowed the name of Elephant upon the Point, to mark its vicinity and the place where the fossils were found; and upon the river that of Buckland, in compliment to Dr. Buckland, the Professor of Geology at Oxford, to whom

4

I am much indebted for the arrangement of the geological memoranda attached to this work. In consequence of the shallow water, there was much difficulty in embarking the fossils, the tusks in particular, the largest of which weighed 160 lbs., and it took the greater part of the night to accomplish it." [1]

On his return, Captain Beechey writes to Buckland from Harley Street in October 1828, "The bones arrived yesterday in good order at the Admiralty," and begs him to come with all speed and unpack them, "as the 'cases' are very large and occupy the Hall." The most perfect series was selected for the British Museum ; [2] another series, including some of the largest tusks of elephants, was sent to the Museum at Edinburgh; others to the Geological Society of London.

Another Arctic explorer, who was Buckland's old and valued friend, was Sir John Franklin. After his return from his second voyage to the Arctic regions, he came to Oxford as the hero of the day to receive the honorary degree of D.C.L. On this occasion he and his daughter were the guests of Buckland at Christ Church. Always taking the keenest interest in Arctic discoveries, Buckland was one of Lady Franklin's chief advisers in the several expeditions organised to search for the lost explorers. Subsequently both Sir Leopold M'Clintock and Admiral Inglefield were frequent guests at the Deanery of Westminster.

---

[1] Beechey's "Voyage to the Pacific."
[2] Now to be seen in the Natural History Museum, Cromwell Road, South-East Gallery, Ground Floor, Cases 10, 16, and 31.

During Buckland's lifetime his collection was placed in the Clarendon Buildings. A room had been prepared for its reception, and the Professor writes in the highest glee to Lady Mary Cole, on April 3rd, 1822, to tell her of the fact. "You will be pleased to hear that my Lecture Room is to be put to rights and fitted up with £300 worth of cabinets between this and midsummer, when Mr. Miller of Bristol is to come here, and arrange and catalogue my collection, and clear my room of boxes."

Buckland was particularly careful to put descriptive labels on all specimens that came into his possession, and these were usually written, or rather painted, by his wife. From long practice, she acquired a knack of finding the best place on which to mark them, and her clear labelling may be seen on specimens in all parts of the Oxford Museum as well as in Cromwell Road.

Ultimately Buckland bequeathed the collection to the Vice-Chancellor of the University of Oxford for the use of the Professors of Geology who might succeed him, with all the geological charts, sections, and engravings that might be in the Clarendon Buildings at the time of his death. Professor Phillips, who acted as deputy Reader during Dean Buckland's last illness, and succeeded to his Chair, proposed that the collection should henceforth be known by the name of the Bucklandean Museum. A new building was erected about 1858, to which the collection was removed, and a marble bust (by Weekes) of Buckland, the founder of the Geological Collection, was erected by his friends and admirers :—

Marmor hoc egregii Viri

GULIELMI BUCKLAND, S.T.P.,

Soc. Reg. Lond. Soc., Gall. Inst. Sod., Adscr.

in Ecclesia Westmonastr. Decani

Geologiæ apud Oxonienses Professoris insignissimi

oris atque animi lineamenta referens

effingi curaverunt

et in hac Æde studiis naturalibus

dicata

inter longinquæ vetus vetustatis rudera

antiquitatis primitias

ope sua et industria excavatas

in perpetuum conservari voluerunt

Amici et Discipuli superstites.

A.D. 1860.

The subsequent history of the collection is a melancholy record of neglect. Owing to a variety of causes,
a great part of this valuable bequest to the University
remains in the same condition (and with perishing labels)
in which it was removed from the Clarendon thirty-six
years ago. The Hebdomadal Council at Oxford were
urged to apportion a space, when the enlargement of the
Museum buildings was contemplated, for the "collection in the cellars," as it was called, and within
the last two years a large room has been placed at the
disposal of the Professor of Geology. There the matter
rests, and, it is feared, will continue to rest, unless the
University makes a special grant to rescue this bequest
from oblivion. Not only does this collection consist of
Dr. Buckland's gatherings of the first-fruits of the new
science, but, as he was the greatest authority on geology
at the beginning of the century, it includes specimens

sent him from all parts of the world. Fortunately for science, Dr. Buckland sent duplicate specimens to the British Museum.

Professor Boyd Dawkins [1] writes respecting this once famous collection :—

"In 1857 Dr. Buckland's collection was in the old Clarendon Buildings, partly in upright glass cases and partly in drawers below. Professor Phillips let me have the run of them, and I spent a good deal of time in working at them ; they were all accessible and were mostly unpacked. They were removed to the new Museum, and the arrangement disturbed, so that at present the collection is in a state unworthy of Oxford. The Bucklandean tradition and name, which were maintained in Oxford down to the death of Phillips, are now almost unknown. The Bucklandean collections are now scarcely known as such."

Among the most interesting, and to his class most familiar, specimens which his collection contained, was the skull of a hyena.

"In the Oxford Museum," says Frank Buckland, "is a very perfect skull of one of our ancient British Cave Hyenas ; and my father, in his usual clever manner, often made it appear in his lectures (and with good reason too) that this skull was that of the old cannibal, Paterfamilias of his cave, who devoured and survived all his relations. The following verses were composed by one of the class upon 'The Last English Hyena':—

" 'High on a rock, which o'er the raging flood
    Reared its bleak crag, *The Last Hyena* stood.
    Beneath his paws a kindred skull was seen ;
    And he, with commons short, looked grim and lean.

---

[1] W. Boyd Dawkins, M.A. (Oxon.), F.S.A., F.G.S., Professor of Geology and Palæontology in the Victoria University, Owens Coll., Manchester.

" 'Potent his jaw to crack his bony rapine,
Potent his stomach as a " pot of Pappin ";
O'er this last bone of many a murdered brother
He growled, for he in vain had sought another.

" 'Full oft, like Captain Franklin, did he prey
On bones neglected on a former day ;
But now th' o'erwhelming surge had buried all,
In caves below, of beasts both great and small.

" 'But ere it rose to mix him with the rest,
Thus did he growl aloud his last bequest :—
"My skull to William Buckland I bequeath."
He moaned—and ocean's wave he sank beneath.

" 'Southward the flood from Yorkshire chanced to travel
And rolled the monster deep in Yorkshire gravel.
Behold the head of that Hyena grim,
Who through *Diluvium* deeps essayed to swim.'

" After vast labour and much accurate observation, Dr.
Buckland at length made the evidence of the former
existence of hyenas in England quite complete ; so complete indeed, that on one occasion, when surrounded by
the actual bones and specimens knocked out of the
Kirkdale stalactite by his own hammer, and brought to
Oxford by his own hand, and sitting in his Professor's
chair in his own museum, he appealed to one of the most
learned judges of the land, who happened to be present
at his lecture. After having, with his usual forcible and
telling eloquence, put his case, to prove not only the former
existence of hyenas in England, but even that they were
rapacious, ravenous, and murderous cannibals, he turned
round to the learned lawyer and said, ' And now, what do
you think of that, my lord ? ' ' Such facts,' replied the
judge, ' brought as evidence against a *man*, would be quite
sufficient to convict and even hang him.' "[1]

[1] F. Buckland, "Curiosities of Natural History," 2nd series, pp. 52, 53.

CHAPTER III.

BONE CAVES AT KIRKDALE, GOAT'S HOLE, DREAM LEAD MINE, AND GAILENREUTH; PUBLICATION OF "RE-LIQUIÆ DILUVIANÆ," 1823; FIRST PRESIDENT OF THE ROYAL GEOLOGICAL SOCIETY, 1824.

1822—1824.

IN 1822 Dr. Buckland addressed to the Royal Society, of which he had been elected Fellow in 1818, a paper describing his researches in the bone cave of Kirkdale, which had been discovered in the preceding year in the Vale of Pickering, about twenty-five miles from York, and was the first fossil cave known in England. This paper was printed in the Philosophical Transactions for 1822, and made so considerable an impression that its author was, in the same year, honoured with the Copley Medal of the Royal Society.

In 1823 Dr. Buckland published in a quarto volume his "Reliquiæ Diluvianæ." This important work made his name still more widely known. To its value and influence Professor Boyd Dawkins of Owens College, Manchester, bears warm testimony:—

"I give you the impression of one who as an undergraduate fell under the influence of Dr. Buckland's name handed down by tradition at Oxford, and who afterwards,

as a teacher, has had some experience of the value of his
work.   When I went up to Oxford in 1857, Dr. Buckland's
name was a great memory in the University, and Professor
Phillips, who had worked with him side by side almost
from the beginning of his geological work, was giving
lectures in the Old Clarendon Buildings facing the Broad.
The parts of these lectures which left a lasting mark on
me were those relating to the liassic reptiles, and the
Reptiles and Mammalia from Stonesfield, both of which
had been either discovered, or specially dealt with, by Dr.
Buckland in the Bridgewater Treatise, and were in his
collection.   There were also the large collections from the
bone caves of Germany and England described by him in
the ' Reliquiæ Diluvianæ,' which profoundly impressed me
and caused me to take up more particularly that section
of the history of the earth about which I have written in
  Cave Hunting.'
   " I therefore in my own person can speak of the great
influence which Dr. Buckland's work has had on me,
either directly from his collection, or through his friend
Professor Phillips.   I shall never cease to venerate his
name.   His books still, in my opinion, belong to the classics
of Geology, although of course during the last seventy
years the theories as to the Deluge and the doctrine of
Final Causes have changed.   The facts, however, have not
changed, and, in my work as Professor in Owens College, I
still use as a class-book the last edition of the Bridgewater
Treatise edited by Phillips for the Reptiles, the Stonesfield
Mammalia, and the Pentacrinoids.   My book on ' Cave
Hunting' is a lineal descendant of ' Reliquiæ Diluvianæ,'
and probably I should never have taken up that question
had Dr. Buckland's book never been written."

   When Dr. Buckland was writing this essay on Kirkdale
Cavern, he took great pains to compare the bones there
found with recent bones, in order to make his story quite
complete.

   " In the Kirkdale cave he found a portion of a skull

which he believed belonged to a young hyena, and although
nearly certain that it was what he thought it to be, he
ransacked all the collections he knew for a recent skull
of a young animal for comparison ; and not finding one, he
requested Mr. Burchell, the great African traveller, to send
him a young hyena from the Cape. In course of time the
baby-beast arrived in the Docks ; a pretty tame little beast,
a great favourite with the sailors, who had christened him
' Billy,' doomed nevertheless to be slain for the sake of
science. The late Mr. Cross, then of Exeter 'Change, and

KIRKDALE CAVE.

afterwards of the Surrey Zoological Gardens, acted as agent,
and undertook the delivery of poor ' Billy.' The little brute,
however, by his good temper and playful manners, quite
won the heart of Mr. Cross, who begged hard for his life,
and at length obtained a respite on the condition that
the skull of a young hyena should be forthcoming. Mr.
Cross, we suppose, turned out all his drawers and cabinets
in search ; anyhow, he, within the given time, produced
a skull, which was not the skull of poor Billy. His life
was spared, and he was forthwith taken to Exeter 'Change,
and thence removed with the rest of the wild beasts to the
Surrey Zoological Gardens.

" ' The Companion to the Royal Menagerie, Exeter 'Change, containing concise descriptions, scientific and interesting, of the curious foreign animals now in that eminent collection, derived from actual observation, by Edward Cross, Proprietor, 1820,' describes Billy, then in his youth, but amiable withal :—' The hyena in a cage at the end of the room is possessed of a large share of good humour, and entertains the visitors at feeding time by the gesticulations of delight he manifests at the moment, and by his curious imitations of the human voice resembling laughter. This animal suffers himself to be caressed, and is so familiar with the keepers, that when any repairs are wanting in his cage they have no hesitation in going in with him. (N.B.—This was before the day of Van Hamborough, and other lion kings.) He is a native of the Cape of Good Hope, and is frequently called the Tiger Wolf.' Billy arrived in England in the year 1820, and he died in his den a peaceable quiet death, January 14th, 1846, having lived just a quarter of a century within this metropolis. . . .

" At his decease (the cause of death, *plus* old age, being an enormous goitre in the throat), Dr. Buckland presented his carcass to the Royal College of Surgeons, reserving, however, the skin for himself. . . . Billy first made his *début* as the youngest hyena in England ; he ended his career grim and grisly as the oldest hyena in England, and probably in Europe. The stuffed skin is now at the College of Surgeons in company with his skeleton, having been bought at the sale (of the Dean's effects, January 1857) by Professor Quekett. Not only was Billy subservient to the cause of science when dead, but even when alive he unknowingly gave much important assistance to his former owner, then busy with the ' Reliquiæ Diluvianæ,' for Billy cracked the marrow-bones of oxen, and refused those bones which contained no marrow, exactly as did his ancestors ages before him in the wilds of Yorkshire, as yet untrodden by the foot of man. So wonderfully alike were these bones in their fracture, that, judging from this point alone, it was impossible to say

which bone had been cracked by Billy and which by the
aboriginal hyena of Kirkdale.   Again, Billy polished with
his feet and hide the sides and floor of his den of wood
as his ancestors did the sides and floor of their den of
stalactite in the Yorkshire hills ; and as the ancient beasts
deposited *album græcum* in abundance after a dinner of
bones, so did Billy deposit pounds of the same substance,
even in this minute circumstance illustrating the history
of his ancient British forefathers." [1]

No one then believed either in the probability or
possibility of wild beasts, which now exist only in warm
climates, having lived and died in our Yorkshire wolds.
Hence Dr. Buckland was bound to give proof of his
assertions, and, as usual, spared no pains or trouble in
verifying the novel and extraordinary results of his
examination of the cave.   He took Sir Humphry Davy
to visit it, and writes to the Rev. W. Vernon Harcourt that
the eminent scientist " is satisfied with the accuracy of my
facts."   He adds : " We have had this week in Oxford
a Cape hyena who has performed admirably on shins
of beef, leaving precisely those parts which are left at
Kirkdale and devouring what are there wanting, and
leaving splinters and scanty marks of his teeth on the
residuary fragments which are not distinguished from
those in the den."

Dr. Buckland's interest in hyenas caused some amusement
to his friend Lyell, who writes to Dr. Mantell, in 1826 :
" Buckland has got a letter from India about modern
hyenas, whose manners, habitations, diet, etc., are every-

---

[1] Frank Buckland's "Curiosities of Natural History," 2nd series.

thing he could wish and as much as could be expected
had they attended regularly this course of his lectures."

Buckland found in the Kirkdale cave not only remains
of hyenas, but teeth and bones of twenty-three different
animals—among them tiger, bear, wolf, elephant, rhinoceros, hippopotamus, horse, ox, three species of deer, hare,
rabbit, water-rat, mouse. Of birds' remains, he also
found raven, pigeon, lark, snipe, and a small species of
duck resembling the *anas sponsor* or summer duck. This
wonderful cave no longer exists, having been quarried
away. Buckland says : " The workmen on first discover-.
ing the bones at Kirkdale cave supposed them to have
belonged to cattle that died of a murrain in this district
a few years before, and they were for some time neglected
and thrown on the roads with the common limestone."
It was to the kindness of the Bishop of Oxford (Legge)
that the Professor was indebted for the first information
as to the existence of the cave. He visited it in December
1821, and described the entrance as "a hole in the perpendicular face of the quarry about three feet high and five
feet broad, which it is only possible for a man to enter on
his hands and knees, and which expands and contracts
itself irregularly, from two to seven feet in breadth and
two to fourteen feet in height, diminishing however as it
proceeds into the interior of the hill."

From Dr. Buckland's minute account of the contents of
the cave the following abridged extract may be taken :—

" The bottom of the cave, on first removing the mud, was
found to be strewed all over like a dog kennel, from one
end to the other, with hundreds of teeth and bones, or

rather broken and splintered fragments of bones, of all the animals above enumerated; they were found in greatest quantity near its mouth, simply because its area in this part was most capacious; those of the larger animals, elephant, rhinoceros, etc., were found co-extensively with all the rest even in the inmost and smallest recesses. Scarcely a single bone has escaped fracture, and on

BUCKLAND ENTERING THE KIRKDALE CAVERN. FROM A CARICATURE BY THE REV. W. CONYBEARE.

some of the bones, marks may be traced, which, on applying one to the other, appear exactly to fit the form of the canine teeth of the hyena that occur in the cave. The hyena's bones have been broken, and apparently gnawed equally with those of the other animals. Not one skull is to be found entire; fragments of jaw-bones are by no means common; the ordinary fate of the jaw-bones, as of all the rest, appears to have been to be broken to

pieces and swallowed, the teeth being rejected as too hard
for mastication, and without marrow. The greatest
number of teeth are those of hyenas and the ruminantia.
Mr. Gibson alone collected more than three hundred canine
teeth of the hyena, which at the least must have belonged
to seventy-five individuals, and adding to these the canine
teeth I have seen in other collections, I cannot calculate
the total number of hyenas of which there is evidence at
less than two hundred or three hundred. The only remains
that have been found of the tiger species are two large
canine teeth and a few molar teeth. There is one tusk
only of a bear, which exactly resembles those of the extinct
Ursus spelæus of the caves of Germany, the size of which,
M. Cuvier says, must have equalled those of a large horse.
It is probable that the cave at Kirkdale was, during a long
succession of years, inhabited as a den by hyenas, and that
they dragged into its recesses the other animal bodies
whose remains are found mixed indiscriminately with their
own." [1]

Buckland's friend the Rev. William Conybeare made
a caricature of the Professor entering the cave, and wrote
the following amusing verses :—

> " But of all the miraculous caves,
>   And of all their miraculous stories,
> Kirby Hole all.its brethren outbraves,
> With Buckland to tell of its glories.

> "Ages long ere this planet was formed,
>   (I beg pardon—before it was drowned,)
> Fierce and fell were the monsters that swarmed,
> Roared, and rolled in these hollows profound.

> "I can show you the fragments half-gnawed,
>   Their own *album græcum* I've spied,
> And here are the bones that they pawed,
> And polished in scratching their hide.

----

[1] " Reliquiæ Diluvianæ," p. 15.

"I know how they fared every day,
  Can tell Sunday's from Saturday's dinner;
What rats they devoured, can say,
  When the game of the forest grew thinner.

"Your elk of the bogs was a meat
  That each common hunt might obtain,
But an elephant's haunch was a treat
  They only could hope now and then.

"Mystic cavern! the gloom of thy cell,
  Shedding light on each point that was dark,
Tells the hour by Shrewsbury clock
  When Noah went into the ark.

"By the crust on the stalactite floor,
  The post-Adamite ages I've reckoned—
Summed their years, days, and hours, and more,
  And find it comes right to a second.

"Mystic cavern! thy clearness sublime
  All the chasms of history supply;
What was done ere the birthday of Time,
  Through one other such hole I could spy."

Another famous cave was Paviland or Goat Hole, in
the district of Gower. This discovery is on the coast of
Glamorganshire, fifteen miles west of Swansea, between
Oxwich Bay and the Worm's Head, on the property of
C. M. Talbot, Esq., of Penrice Castle. It consists of two
large caves facing the sea in the front of a lofty cliff of
limestone, which rises more than one hundred feet perpendicularly above the mouth of the caves, and below them
slopes at an angle of 40° to the water's edge, presenting
the bluff and ragged shores to the waves, which are very
violent along this north coast of the estuary of the
Severn. These caves are altogether invisible from the

land side, and are accessible only at low water, except by
dangerous climbing along the face of a nearly precipitous
cliff, composed entirely of compact mountain limestone.

"One of them only called Goat's Hole," writes Buckland in the 'Reliquiæ Diluvianæ,' "had been noticed when
I arrived there. . . . Its existence had been long known
to the farmers of the adjacent lands, as well as the fact
of its containing large bones, but it had been no further
attended to till last summer, when it was explored by the
surgeon and curate of the nearest village, Porteynon, who
discovered in it two molar teeth of elephant and a
portion of a large curved tusk, which latter they buried
again in the earth, where it remained till it was extracted
a second time, on a further examination of the cave by
L. W. Dillwyn, Esq., and Miss Talbot, and removed to
Penrice Castle, together with a large part of the skull to
which it had belonged, and several baskets full of other
teeth and bones. On the news of this further discovery being
communicated to me, I went immediately from Derbyshire
to Wales, and found the position of the cave to be such as
I have above described ; and its floor at the mouth to be
from thirty to forty feet above high-water mark, so that
the waves of the highest storms occasionally dash into it,
and have produced three or four deep rock basins in its
very threshold, by the rolling on their axis of large stones,
which still lie at the bottom of these basins ; around their
edge, and in the outer part of the cave itself, are strewed
a considerable number of sea pebbles, resting on the native
limestone rock. . . . Where the pebbles cease, the floor is
covered with a mass of diluvial loam of a reddish yellow
colour, abundantly mixed with angular fragments of limestone and broken calcareous spar, and interspersed with
recent sea shells, and with teeth and bones of the following
animals, viz. elephant, rhinoceros, bear, hyena, wolf, fox,
horse, ox, deer of two or three species, water-rats, sheep,
birds, and man.

"I found also fragments of charcoal, and a small flint,

the edges of which had been chipped off, as if by striking
a light.[1] . . .

"In another part I discovered beneath a shallow covering of six inches of earth nearly the entire left side of
a human female skeleton. The skull and vertebræ, and
extremities of the right side were wanting ; the remaining
parts lay extended in the usual position of burial and in
their natural order of contact. In the middle of the bones
of the ancle was a small quantity of yellow wax-like substance resembling adipocere. All the bones were stained
superficially with a dark brick-red colour, and enveloped
by a coating of a kind of ruddle, which stained the earth,
and in some parts extended itself to the distance of about
half an inch around the surface of the bones. The body
must have been entirely surrounded or covered over at the
time of its interment with this red substance. Close to
that part of the thigh bone where the pocket is usually
worn, I found laid together and surrounded also by
ruddle about two handsful of small shells of the *Nerita
littoralis* in a state of great decay, and falling to dust on
the slightest pressure. At another part of the skeleton,
viz. in contact with the ribs, I found forty or fifty fragments
of small ivory rods. . . . In another place were found
three fragments of the same ivory, which had been
cut into unmeaning forms by a rough-edged instrument,
probably a coarse knife, the marks of which remain on
all their surfaces. One of these fragments is nearly of
the shape and size of a human tongue. No metallic
instruments have been as yet found amongst these remains,
which, though clearly not coeval with the antediluvian
bones of the extinct species, appear to have lain there
many centuries. The charcoal and fragments of recent
bone that are apparently the remains of human food,
render it probable that this exposed and solitary cave has
at some time or other been the scene of human habitation,

---

[1] Dr. Buckland states that the most remarkable of the remains of
these animals are preserved in the collection at Penrice Castle, and in
the museum at Oxford.

if to no other persons, at least to the woman whose bones I have been describing.

"The ivory rings and rods and tongue-shaped fragment are certainly made from part of the antediluvian tusks that lay in the same cave ; and as they must have been cut to their present shape at a time when the ivory was hard, and not crumbling to pieces as it is at present on the slightest touch, we may assume to them very high antiquity, which is further confirmed by the decayed state of the shells that lay in contact with the thigh bone, and, like the rods and rings, must have been buried with the woman. The circumstance of the remains of a British camp existing on the hill immediately above this cave, seems to throw much light on the character and date of the woman ; and whatever may have been her occupation, the vicinity of a camp would afford a motive for residence as well as the means of subsistence in what is now so exposed and uninviting a solitude.

"The fragments of charcoal and recent bones of oxen, sheep, and pigs, are probably the remains of culinary operations; the larger shells may have been collected, also for food, from the adjacent shore, and the small nerite shells either have been kept in the pocket for the beauty of their yellow colour, or have been used, as I am informed by the Rev. Henry Knight, of Newton Nottage, they now are in that part of Glamorganshire, in some simple species of game. The ivory rods also may have been applicable to some game, as we use chess men or pins of a cribbage board ; or they may be fragments of pins, such as Sir Richard Hoare has found in the barrows of Wilts and Dorset, together with large bodkins also of ivory, and which were probably used to fasten together the coarse garments of the ancient Britons. It is a curious coincidence also, that he has found in a barrow near Warminster, at Cop Head Hill, the shell of a nerite and some ivory beads, which were laid by the skeletons of an infant and an adult female, apparently its mother.

"That ivory rings were at that time used as armlets, is probable from the circumstance of similar rings having

also been found by Sir Richard Hoare in these same
barrows ; and from a passage in Strabo, lib. 4, which Mr.
Knight has pointed out to me, in which, speaking of the
small taxes which it was possible to levy on the Britons,
he specifies their imports to be very insignificant, consisting chiefly of ivory armlets and necklaces, Ligurian stones,
glass vessels, and other suchlike trifles.   The custom of

Plan.

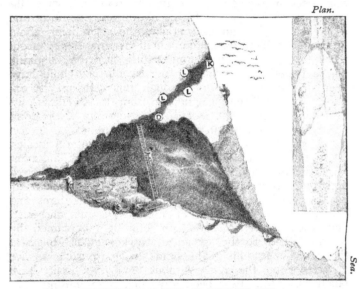

SECTION OF GOAT HOLE, OR PAVILAND CAVE.

burying with their possessors the ornaments and chief
utensils of the deceased, is evident from the remains of
this kind discovered everywhere in the ancient barrows ;
and this may explain the circumstance of our finding with
the bones of the woman at Paviland the ivory rods, and
rings, and nerite shells, which she had probably made use
of during life.   I am at a loss to conjecture what could
have been the object of collecting the red oxide of iron
that seems to have been thrown over the body when laid

in the grave : it is a substance, however, which occurs
abundantly in the limestone rocks of the neighbourhood.
From all these circumstances there is reason to conclude
that the date of these human bones is coeval with that of the
military occupation of the adjacent summits, and anterior to,
or coeval with, the Roman invasion of this country. . . .

"It remains only to describe a long, cavernous aperture
that rises like a crooked chimney from its roof to the
nearly vertical face of the rock above : its form and
diameter are throughout irregular, the latter being about
twelve feet where longest, and in its narrowest part about
three feet ; so that it is impossible the large elephant,
whose bones were found in the cave below, could have
been drifted down entire through this aperture. It expands and contracts irregularly from D (see Plate), its
lower extremity in the roof of the cavern, to K, the point
at which it terminates in the face of the cliff.

"Along this tortuous ascent are several lateral cavities,
L.L.L. the bottoms of which afford a place of lodgment
for a bed of brown earth, about a foot thick, and derived
apparently from dust driven in continually by the wind.
In this earth I found the bones of various birds and
fish, and a few land shells, of moles, water-rats, and mice,
and their presence here can only be explained by
referring them to the agency of hawks, and fish-bones to
that of the seagulls. The land shells are such as live
at present on the rock without, and may easily have
fallen in. Had there been any stalagmite uniting these
bones into a breccia[1] they would have afforded a perfect analogy to the accumulation of modern birds' bones,
by the agency of hawks, at Gibraltar ; where Major Imrie
describes them as forming a breccia of modern origin in
fissures of the same rock which has other cavities filled
with a bony breccia of more ancient date, and which are
of the same antediluvian origin with the older parts of the
bones that occur on the floor of the cave at Paviland."[2]

---

[1] Breccia consists of fragments of different rocks cemented together.
[2] Miss Talbot, the present owner of Penrice Castle, writes from

The following *jeu d'esprit* on the female skeleton found by Dr. Buckland in the Paviland Cave is from the pen of Mr. Philip Duncan :—

> " Have ye heard of the woman so long underground ?
> Have ye heard of the woman that Buckland has found,
> With her bones of empyreal hue ?
> Oh, fair one of modern days! hang down your head,
> The antediluvians rouged when dead—
> Only granted in lifetime to you ! "

A third cave was that which was discovered in the Dream Lead Mine, Derbyshire. The lead mine called the Dream is in the hamlet of Caelow, about one mile from Wirksworth, and on the property of Philip Gell, Esq., by whose exertions nearly the entire skeleton of a rhinoceros was extracted, together with some considerable remains of the horse, ox, and deer.

Buckland thus describes the cave and its contents :—

" In December 1822, some miners engaged in pursuing a lead vein had sunk a shaft about sixty feet through solid mountain limestone, when they suddenly penetrated a large cavern, filled entirely to the roof with a confused mass of clay and fragments of stone, through which they attempted to continue their shaft perpendicularly downwards to the vein below ; in this operation they were interrupted by the earth and fragments beginning to move and fall in upon them continually from the sides until the roof of a large cavern became apparent. It was nearly in the centre of this subsiding mass, and at the height of many feet above the floor of the cave, that the workmen discovered the bones of a rhinoceros. They lay very near to each other, and probably formed an entire skeleton before they were disturbed.

Margam December 13th, 1892 : " We have a number of bones at Penrice Castle which were found in Paviland Cave, but I believe the bulk of them were taken to the Swansea Museum."

The bones are in a state of high preservation, and from a nearly full grown animal, and, being found so close together, are without doubt portions of a skeleton which lay entire in the middle of the cave before the materials that had filled it began to subside. There were no supernumerary bones, to indicate the presence of a second rhinoceros; but in the same cave were found some teeth and bones of a horse, and many entire bones from the legs of a very large ox, all apparently from one individual; also many bones of deer from at least four individuals, and fragments of horns, none so large as those of red deer. From the circumstance that none of these bones have marks of partial decay on one surface only, we may infer that they were derived from animals that perished by the waters that introduced them to the cave: they are of a yellowish brown colour. . . .

"For some time after the cave was penetrated there was no apparent communication between its interior and the surface; but as the loose materials that at first filled it subsided into and were taken out by the shaft, a sinking appeared in the field above at I, and a further mass of the same kind, viz. clay and fragments of limestone, mixed with a few rolled pebbles of quartz, continued to fall downwards into it (like the contents of a limekiln, sinking towards the lower aperture by which the lime is extracted), until a large open chasm D, more than six feet broad, and fifty feet deep, was left entirely void, and seemed to form .a direct communication from the side of the cave to the surface of the field above. Till undermined in this manner, the fissure D had been entirely filled, and the surface afforded not the slightest indication of its existence; at present it is restored to the same state of an open chasm in which it probably was before the access of the diluvian waters, that appear to have swept into it the mud and rocky fragments which filled both it and the cave below; and on examining its sides, I found the projecting parts of them rubbed and scratched by the descent of these heavy. bodies as they dropped in from above.

"From the situation of the rhinoceros' bones in the middle of this drifted mass, and in the centre of the cave,

added to the juxtaposition of so many of the component
parts of one entire skeleton, which are neither rolled, or
gnawed, or broken, except by the movement they have
recently undergone, and the pickaxes of the miners, it
seems probable that they are the remains of a carcase that
was drifted in entire at the same time with the diluvial
detritus, in the midst of which they were found embedded :

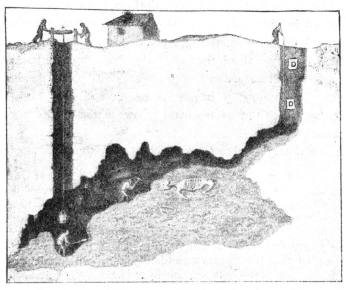

SECTION OF CAVE IN DREAM LEAD MINE, NEAR WIRKSWORTH, DERBYSHIRE.

had they been washed in singly, they would have been
slightly rolled and scattered irregularly, and we should
probably have found parts of more than a single individual ;
and had they been derived from an animal that fell into
the fissure, and perished before the introduction of the
diluvium, they would not have been suspended, as they
were, altogether nearly in the middle of it, but would have
lain either on the actual floor of the cave beneath the
loam and pebbles, or have been scattered and drifted

irregularly to different and distant parts of its lowest recesses. I could discover no stalagmite [1] and but few traces of stalactite in any part of this cavern, or of the fissure immediately connected with it." [2]

Although Buckland describes several German caves in his "Reliquiæ Diluvianæ," it will suffice to select the cave of Gailenreuth, near Muggenendorf, in Bavaria, which he visited in 1816, and again in 1822. It is by far the most remarkable cave in Germany, both for the quantity and high preservation of the bones that have been extracted from it, and, like other foreign caves, differs from those of our own country by having its mouth still open, and in the appearance of having been inhabited also in the post-diluvian period.

Buckland describes the Gailenreuth cavern as

"situated in a perpendicular rock, in the highest part of the cliffs which form the left side of the valley of the Weissent River, at an elevation of more than three hundred feet above its bed. . . . The cave consists principally of two large chambers, varying in breadth from ten to thirty feet, and in height from three to twenty feet: the roof is in most parts abundantly hung with stalactite; and in the first chamber, the floor is nearly covered with

---

[1] Stalactites are like icicles of stone hanging from the roofs of caverns, formed by the dropping of water containing particles of lime through fissures and pores of rocks. Stalagmites are a deposit of stalactitic matter on the floors of caverns, sometimes rising into columns which meet and blend with the stalactites above.

[2] "Reliquiæ Diluvianæ," pp. 61 to 64. The bones from the caves of Gailenreuth and Kirkdale, and a bit of the red woman's bone from Paviland, can be seen beautifully arranged in a case on the right hand side of the Geological Gallery of the Natural History Museum at South Kensington.

stalagmite, piled in irregular mamillated heaps, one of
which in the centre is accumulated into a large pillar
uniting the roof to the floor. We descend by ladders
to a second chamber, the floor of which also appears to
have been once overspread with a similar stalagmitic
crust : this, however, has been nearly destroyed by holes
dug through it, in search of the prodigious quantities of

VERTICAL SECTION OF THE CAVERN AT GAILENREUTH IN FRANCONIA.

bones that lie beneath. The cave is connected by a low
and narrow passage, with a smaller cavern, at the bottom
of which is a nearly circular hole, descending like a well
about twenty-five feet, and from three to four feet in
diameter, into which you let yourself down, as in climbing
a chimney, by supporting the hands, feet, and back
against the opposite sides.   The circumference of this
hole is for the most part composed of a breccia of bones,
pebbles, and loam, cemented by stalagmite : on one side of

it is a lateral cavity, which is entirely artificial, and is the spot from which the most perfect skulls and bones have been extracted in the greatest abundance ; the lowest cavity is also entirely surrounded with the breccia above described. The roof and the sides of the artificial cavities, having been dug in the breccia, are crowded with teeth and bones ; but these latter do not occur in the roof or sides of any of the upper or natural chambers above the level of the stalagmitic crust that covers their floor ; this applies equally to all the other caverns I have been describing."

One more passage may be quoted from " Reliquiæ Diluvianæ." The passage describes the Siberian mammoth (*Mammoth Elephas Primigenius*) preserved in the Museum of the Academy of Sciences, St. Petersburg. The specimen has much of the dead skin still covering the head and feet. Its carcass was originally found entire, buried in frozen mud near the mouth of the river Lena, in Siberia, and the skeleton was brought to St. Petersburg by Adams in 1806. A portion of its skin and hair was presented to Dr. Buckland, and he esteemed this relic as one of his greatest treasures. In 1825 he writes to the Rev. W. Vernon Harcourt, telling him that the Bishop of Durham, Shute Barrington, to whom he had dedicated the " Reliquiæ Diluvianæ,"

" has required me to have the lock of hair, etc., of the Siberian mammoth preserved in some appropriate manner at his expense. I mean to place it under crystal in the cover of a box of fossil ivory, if I can get any sufficiently hard, which I have not here ; but I remember that a lapidary and curiosity collector at Burlington, whose name you probably know, but I forget, has just such a piece of a tusk from that coast. A great part of it had actually been made into boxes, and the remainder was in his collection, being four or five inches long, and for which he asked a very high price, four or five guineas. If I can get one and a half inches long from the most perfect end I shall not quarrel with the price."

Eventually these precious relics of the mammoth were enshrined in a silver box. In the "Reliquiæ Diluvianæ" the monster itself is thus described :—

"The fossil elephant differs from any living species of that genus, but approaches more nearly to the Asiatic than to that of Africa. The term mammoth (animal of the earth) has been applied to it by the natives of Siberia, who imagined the bones to be those of some huge animal that lived at present like a mole beneath the surface of the earth. It appears from the wonderful specimen that was found entire in the ice of Tungusia, that this species was clothed with coarse tufty wool of a reddish colour, interspersed with stiff black hair, unlike that of any known animal ; that it had a long mane on its neck and back, and had its ears protected by tufts of hair, and was at least sixteen feet high.

"The bones of elephants occurring in Britain had from very ancient times attracted attention, and are mentioned with wonder by the early historians. The old and vulgar notion that they were gigantic bones of the human species is at once refuted by the smallest knowledge of anatomy. The next idea, which long prevailed, and was considered satisfactory by the antiquaries of the last century, was, that they were the remains of elephants imported by the Roman armies. This idea is also refuted : first by the anatomical fact of their belonging to an extinct species of this genus ; secondly, by their being usually accompanied by the bones of rhinoceros and hippopotamus, animals which could never have been attached to Roman armies ; thirdly, by their being found dispersed over Siberia and North America, in equal or even greater abundance than in those parts of Europe which were subjected to the Roman power. The later and still more rational idea,

that they were drifted northwards by the diluvian waters
from tropical regions, must be abandoned on the authority
of the evidence afforded by the den at Kirkdale; and it
now remains only to admit that they must have inhabited
the countries in which their bones are found.

" In the streets of London the teeth and bones are often
found, in digging foundations and sewers, embedded in
the gravel ; *e.g.* elephants' teeth have been found under
twelve feet of gravel in Gray's Inn Lane ; and lately at
thirty feet deep, in digging the grand sewer, near Charles
Street, on the east of Waterloo Place. At Kingsland, near
Hoxton, in 1806, an entire elephant's skull was discovered,
containing two tusks of enormous length, as well as the
grinding-teeth ; they have also been frequently found at
Ilford, on the road from London to Harwich, and, indeed,
in almost all the gravel-pits round London. The teeth
are of all sizes, from the milk-teeth to those of the larger
and most perfect growth ; and some of them show all
the intermediate and peculiar stages of change to which
the teeth of modern elephants are subject. In the gravel-
pits of Oxford and Abingdon, teeth and tusks, and various
bones of the elephant, are found mixed with the bones
of rhinoceros, horse, ox, hog, and several species of deer,
often crowded together in the same pit, and seldom rolled
or rubbed at the edges, although they have not been found
united in entire skeletons.

" In the Ashmolean Museum there are some vertebræ,
and a thigh bone of an enormous elephant, at least sixteen
feet high, which are in the most delicate state of preservation, and were found in the gravel at Abingdon four or five
years ago. In the same pit with them were collected also
fragments of sixteen horns of deer. . . . About three years
since a large molar tooth of an elephant was dug up in a
gravel-pit in one of the streets of Oxford, in front of
St. John's College. . . . At Newnham, in Warwickshire,
near Church Lawford, about two miles west of Rugby, two
magnificent heads and numerous bones and teeth of several
individuals of the Siberian rhinoceros, with many large
tusks and teeth of elephants, and some stags' horns, and

bones of the ox and horse, were found, in the year 1815, in a bed of diluvium, which is immediately incumbent on stratified beds of lias. . . . One of these heads, measuring in length two feet six inches, together with a small tusk, and molar tooth of an elephant, have, by the kindness of Henry Hakewell, Esq. (of architectural celebrity), been deposited in the museum at Oxford." [1]

A curved tusk, from the same place, measuring seven feet in length, together with a highly valuable collection of the bones of rhinoceros,[2] were deposited in the Oxford Museum till Dr. Buckland placed his collection in the Clarendon.

The book achieved a remarkable success. Buckland writes to the Rev. W. Vernon Harcourt, December 3rd, 1823 :—

" I am very proud of the rapid sale my book has had ; not a copy has been left for some time. Mr. Murray is very busy in bringing out a second edition of one thousand copies more. You of course have seen the very flattering review of it in the *Quarterly*—it is by Dr. Copleston."

Later on in the month he says :—

" The second edition of my first volume comes out this week, and Mr. Murray tells me he has already sold four hundred copies of it to the booksellers, and expects the whole edition (one thousand copies) will be out of print in six months. I cannot but think myself very successful in my first attempt at a quarto vol.

" I have just been to London to sit on a Committee of

---

[1] " Reliquiæ Diluvianæ," pp. 174 to 177.

[2] Many of these latter have been engraved in Cuvier's "Animaux Fossiles," vol. ii., from drawings by Miss Morland, whom Dr. Buckland afterwards married.

the Royal Society for selecting the best granite for the new London Bridge."

In another letter to the same friend he writes, two years later :—

"I have just received a gold snuff-box set with mosaic from the Emperor of Russia, in acknowledgment of a copy I sent him of my 'Reliquiæ Diluvianæ.' I am just returned," he continues, "from exploring two more hyenas' dens in Devonshire. They were less populous than Kirkdale, but have abundance of splinter and a fair supply of toes and teeth. I found the teeth of rhinoceros in addition to hyenas, bears, and tigers, which have been noticed there by Trevelyan, and found also a flint knife of the same kind as the one I have from Paviland, showing both these caves to have been inhabited by people who used such knives, *i.e.* aboriginal Britons. In the other at Chudleigh I delighted Lord Clifford by finding, under a thick crust of virgin stalagmite, bears and hyenas of enormous size, and plenty of splinters and gnawed frag- ments in a bed of mud more than five feet deep and of which I did not reach the bottom. I passed for a conjurer by telling them where the bones would lie before the crust was touched, and the more so as in three subsequent experiments hardly any bones were found ; it was the finest proof possible of the verity of my theory, for it was precisely in the spot where, according to that theory, they ought to occur in the greatest abundance that they were so found, according to the entire convictions and con- version of Lord Clifford, who had before been persuaded by G. Penn to be an unbeliever in my book. Sir T. Acland was with me in my examinations of both these caves, and dug as if he had been member for Cornwall rather than for people who, like my constituents, live above ground."

When the "Reliquiæ Diluvianæ" was published, the lime caverns at Dudley had never been opened ; but as the

subject of bone caves will not call for attention at a later
period of the narrative, it will not be out of place here
to translate an account written of them by a scientific
foreigner, who made Buckland's acquaintance and was in
his company when these caves were illuminated.   The
author, who is a German naturalist,[1] thus tells his story :—

"These lime caverns, although only the work of art, are
nearly one English mile in length and about a hundred feet
high ; the breadth may be about seventy-five feet. But for
what purpose were such gigantic caverns made by the
hands of man ? The neighbouring iron works use, as
a flux in smelting, great quantities of lime, and Lord
Ward furnishes this requisite from the Dudley Caverns,
which belong to him.   We can form an idea of the
quantity of lime excavated for this purpose, when we
learn that the above-mentioned caverns were begun less
than ten years ago by the excavation of this stone, and
(according to several estimates given to me) the owner
draws from these lime-pits a yearly income of from £15,000
to £20,000.   Another, and a still larger, cave, begun in the
same way, is even now becoming profitable.   Lord Ward,
on the occasion of the meeting of the British Association
at Birmingham 1839 (under the Presidentship of the Rev.
W. Vernon Harcourt), for the convenience of the company,
had this vast subterranean vault most brilliantly lighted.
It must have cost him many hundreds of pounds, for the
number of lights erected was almost incredible, and in
addition, at short intervals throughout the whole extent
of the cave, artificial lights were burning in galleries hewn
out for the purpose.
"In order to reach the entrance, we descended from a
considerable height through an excavation, and boats were
ready to take us along the canal leading to the interior.

---

[1] "Mittheilungen aus dem Reisetagebuche eines deutschen Natur-
forschers Eng." (Basel, 1842.)

As the illumination of the Dudley Caverns is of very rare occurrence, thousands of people flocked from the surrounding country, in spite of the abominable weather, in order to witness the unusual and beautiful spectacle. Naturally only a small part of the crowd could be conveyed by the boats, and the greater part had to proceed by the galleries. Just as the foremost boat, in which I was, passed through the doorway, which had been boldly blasted in the rock, and we had cast the first glance over the immeasurable sea of light and flame, the long vault resounded with thunder crashes and rollings in quick succession. We thought the foundations of the earth were moving, that they were on the point of giving way, that a great geological catastrophe was approaching which would bury unlearned and learned alike in the bowels of the earth, in order to furnish to posterity some materials of comparative anatomy for the precise definition of the organisation of the Adamite creation. But it was only weak man who made this noise ; at the farthest end of the cavern a huge mass of limestone had been blown up by gunpowder in order to show the company how the caverns were begun, and millions of pieces of stone were torn from the bosom of the earth.

" Whilst the more fortunate of the visitors to the cavern pressed on to the interior of the hill comfortably in their boats, and were alone in a position to see the whole of the fairy-like spectacle, the crowd was obliged to twist and jostle each other through the higher passages and paths ; but the moving throng indispensably contributed to the artistic life of the scene. As the floor of the hewn-out galleries is sometimes broad and somewhat steep, we were able to survey completely the moving masses, and watch the most wonderful groups form themselves and vanish again in the manifold lights. Sometimes the people's faces shone in a greyish red light ; sometimes there appeared, alongside and in front of us, a crowd of ghostly corpses pacing the Lower Regions, so pale and ashy did the people in their stony heights seem to us in our little boats. The miners had been obliged to leave at short

distances massive pillars of limestone, to support the
over-lying hill ; this row of columns gave to the long
cavern an artistic effect, and the moving human stream,
alternately hidden by the pillars and re-appearing in
the free spaces, presented a singular appearance. Very
peculiar, also, was the noise produced by the footsteps and
the simultaneous talking of some thousands of people,
and also the oars of many boats, through the re-echoing
arches of the caverns ; indeed so loud was the humming
and buzzing that we could scarcely hear ourselves
speak.

"When we had nearly reached the farther end of the
cavern, the flotilla of boats was ordered to stop and the
foot-passengers stood still. Then Murchison, who had
made these regions an especial subject of his researches,
ascended to a high point, and announced to the assembled
crowd that he would give them a short description of the
geological condition of the surrounding hill. 'Silence,
silence!' sounded from a thousand throats along the cave,
and in a few moments the buzzing noise had ceased,
and in its place reigned the most complete silence.
Murchison's voice had formerly been accustomed to command a regiment, and had lost nothing of its penetrating
power, which under the prevailing circumstances stood
him and his listeners in good stead. In a clear and
concise speech the distinguished Scottish geologist sought
to make us contemplate the conditions of the surrounding
mountain mass, and to give an idea of the peculiarity of
the formation of coal in general.

" It needed a voice of thunder in order to be understood
by all present in that gigantic subterranean dome. But
that the thirst for knowledge of every one might be satisfied,
Buckland went to the gallery, placed himself on a mighty
block of stone, and lectured for more than an hour, he
and his numerous audience being veiled in the wreathing
sulphur smoke, upon the subject already handled by
Murchison, but in so original and humorous a manner
that he held the attention of his listeners in a way seldom
witnessed. Next he sketched an extremely suggestive

6

picture of the origin of the surrounding country, and the primeval plants and animal world which lie buried in it. As is well known, the English have a peculiar love of regarding Nature from a theological point of view, and the celebrated Oxford geologist, as he proved by his last geological work, is no exception to the rule. The immeasurable beds of iron-ore, coal, and limestone which are found in the neighbourhood of Birmingham, lying beside or above one another, and to which man has only to help himself in order to procure for his use the most useful of all metals in a liberal measure, may not, he urged, be considered as mere accident. On the contrary, it in fact expresses the most clear design of Providence to make the inhabitants of the British Isles, by means of this gift, the most powerful and the richest nation on the earth. This theme was treated by Buckland with every permissible variation, to the no small edification of the listening country people, and to my own great pleasure, even though I may not be able to accept his leading idea.

" A rich field for oratorical and humorous development opened itself now, for the first time, as the speaker spoke of the importance of iron. Indeed, where would man be, had not kind fate given him in abundance this plain-looking metal? The possession of this good gift alone enables our race to reach the high level of culture which it holds at the present time. Without it man could never have gained his power over Nature, or enjoyed the immeasurable riches, pleasures, and advantages which the industrial and commercial worlds possess to-day. Without it man's mental horizon would be confined to much narrower limits ; his intellectual development could not have made its present advances, were not iron spread with such a lavish hand over the surface of our own country. When Nature gave this metal to man, she lent him an extraordinary power, bestowed upon him a mighty tool, and raised him from weakness to power, and to the lordship of this earth. If, therefore, there is a providential metal, it is neither gold nor silver, however highly man may estimate them. No, it is iron.

"Let us think, if we were deprived of this metal, what should we be from a physical view? Thousands of bene-fits, thousands of conveniences, which we unconsciously enjoy every hour, would be withdrawn from us ; and how many indispensable necessaries it would be impossible to satisfy! Iron, then, has already become incalculably precious ; its value to the human race has become, in the highest sense of the word, inestimable. Yet still it continues to open out possibilities of immeasurable importance on quite a new side. By its capability of receiving magnetism of extraordinary strength in a moment, and of losing it again in as short a time, iron becomes an inexhaustible source of power. It ministers a mechanical strength to the household, which we can raise according to our inclination and render subject to our rule. It needs no great boldness of imagination to represent the mighty influence which the use of iron has exercised on our social relations. As the magnetic power of iron has for the last century allowed us to find our way across distant seas, so it will, perhaps at no distant period, bring together men by land and sea, bridging over vast spaces with a speed that outstrips the power of steam and vies with the swiftness of the wind.

"I cannot end my comments upon this metal of metals without telling my readers that the excellent Buckland related the history of an old shoe with the most delightful humour. It fell into the hands of an African king, and brought him riches, renown, and respect, owing to there being *nails* in it! Out of this small piece of iron was prepared—I do not know how or of what kind—a tool, which his African majesty lent far and wide for gold dust and other precious things, and through its means greatly raised the amount of his royal revenue.

"After another half-hour's stay underground we gladly sought daylight again, and, amid the singing of 'God save the Queen' from a thousand voices and the thundering crashes of blasted rocks renewed once more, boats and walkers alike left the remarkable vaults of Dudley Caverns."

In 1824 Dr. Buckland secured a Royal Charter for the Geological Society,[1] and was appointed its first President. In February of that year he writes to the Rev. W. Vernon Harcourt an account of the first occasion on which he presided :—

"We had a great meeting in Bedford Street on Friday last, the largest I ever remember. The great attraction was the entire Plesiosaurus which I have purchased for the Duke of Buckingham, and of which Mr. Conybeare on that evening read a description ; the specimen is nearly entire, and, though a young animal, is ten feet long ; when full grown it must have been twenty feet at least. The neck has the very unusual number of forty vertebræ, head like a lizard, neck like a snake, body of a crocodile, paddles like a turtle and two feet long, tail very short, nearly equal to the length of a saddle ; its neck (double as long in proportion as the swan) is an anomaly as yet unique. I had also a paper on the Stonesfield Megalosaurus ; so that with two monsters of such a kind, and so crowded an audience, my first evening of taking the chair as President was one of great *éclat*."

The meetings held at Somerset House were termed "Noctes Geologicæ," and very brilliant these gatherings were when De la Bêche, Conybeare, Smith, Sedgwick, Lyell,[2] Murchison, Owen, Daubeny, Buckland, and others

---

[1] The Geological Society, which was founded on November 13th, 1807, first occupied apartments of its own early in 1809, at No. 4, Garden Court, Temple. In 1816 the Society removed to Bedford Street, Covent Garden, and on April 23rd, 1825, while still in Bedford Street, the Society was incorporated by a Charter, obtained by Buckland (and others), who was at that time President, Charles Lyell being one of the secretaries.

[2] Sir Roderick Murchison says : "If Buckland had done nothing

Rich. Anadell, Pinx.　　　　　　　　　　Walker & Boutall, Ph. Sc.

William Buckland

From a Picture painted about 1843.

met in animated debate. On these occasions Buckland
would draw from his never absent blue bag specimen
after specimen to enforce his arguments ; while his quaint
descriptions would gradually overcome the reserve of his
learned associates and inspire the circle with the geniality
of his temper. One of this select band,—which Sir R.
Murchison loved to call the " Old Guard of Geology,"—the
late Rev. Gilbert Heathcote, Subwarden of Winchester, told
the writer that he well remembered being present on one
of those evenings and hearing Sir Charles Lyell shatter one
of Dr. Buckland's theories. " The veteran laughed heartily
at the demolition of his own theory " by his illustrious
pupil. This is only one of the many instances of
Buckland's large-mindedness and of the heartiness with
which he always welcomed any opinion which seemed a
nearer approach to the truth than he had himself (as a
pioneer) been able to form. Jealousy of those who were
labouring in the same field with himself was entirely foreign
to his nature ; he placed his stores of learning at the
service of others, and writers like Murchison and Agassiz
were indebted to him for the most constant and generous
aid. It was, it may be mentioned, through Buckland's
influence that Murchison adopted the title of " Siluria "
for his book.

The blue bag was an inseparable companion of Buckland's, and it figures largely in all caricatures of the
Professor. " The greatest honour," he used himself to

more than educate a Lyell, a Daubeny, and an Egerton, he would
justly have been placed among the most successful instructors of our
contemporaries."

say, "which my bag ever had was when Lord Grenville insisted on carrying it; and the greatest disgrace it ever had was when I called on Sir Humphry Davy three or four times one day, and always found him out. At last Sir Humphry Davy asked his servant, ' Has Dr. Buckland not called to-day?' 'No, sir; there has been nobody here to-day but a man with a bag, who has been here three or four times, and I always told him you were out.'"

CHAPTER IV.

CANON OF CHRIST CHURCH, JULY 1825; MARRIAGE,
DECEMBER 31ST, 1825; WEDDING TOUR, 1826;
FAMILY LIFE AT CHRIST CHURCH ; MARY ANNING,
THE GEOLOGIST OF LYME REGIS.

1825—1830.

IN 1825 Professor Buckland was presented by his College to the living of Stoke Charity, Hants. In July of the same year he was appointed by Lord Liverpool to a canonry at Christ Church, Oxford, and received the degree of D.D. The appointment necessitated a change of residence. He writes to the Rev. W. Vernon Harcourt on September 10th, 1825 : "Many thanks for your congratulations on my appointment to Christ Church, where I find the hunting of bricklayers and carpenters for the present entirely supersedes that of crocodiles and hyenas."

He carried with him to his new home an enthusiasm for science which was not shared by many of his colleagues. It was not long before he discovered that among the benefactions of Christ Church was one which was available for the promotion of scientific study. Dr. Lee had left a considerable property to the College for a variety of purposes, including the erection and maintenance of an Anatomical

Museum. The property came into the hands of Christ Church in 1766, and in the following year the Museum was built. In the accounts of the Trust entries from time to time occur of purchases of subjects for dissection. But in 1828 Buckland discovered that a considerable sum had accumulated which might be claimed for the benefit of the Museum. In July 1828 he writes to Sir R. Murchison in great delight at his discovery. " I am going to town in a day or two to attend the opening of Brooke's sale, for I have found out £1,200 that we can lay out for our anatomical school at Christ Church, which will quite set us up, unless we find powerful rival bidders in the two new London Colleges." The account books at Christ Church show that large purchases were made. " Dr. Kidd " (then Lee's Reader in Anatomy), " on account of purchases made at Brooke's sale, £500."[1] The smallness of the sum expended compared with that available seems, however, to show that Buckland's fears were realised, and that the Oxford school had found rivals wealthier than themselves.

In this appropriation of a portion of Lee's benefaction to science, Buckland, though neither the Lecturer nor one of the Trustees of the fund, took a leading part, as his letter to Murchison clearly proves. Nor was this the only direction in which he used his new position for the advance of scientific research. He was prompt to avail himself of the services of the masons employed at work on the old Residence attached to the Canonry to make

---

[1] Communicated by the Ven. Archdeacon Palmer.

receptacles in slabs of stone for experimenting on toads. The house which he occupied has since been assigned as the residence of the Archdeacon of Oxford, and Archdeacon Palmer, who now resides there, has placed the stones in a rockery in his garden in memory of these experiments.

" In consequence," Buckland writes, " of many stories of toads being found alive in stones I began in 1825 a series of careful experiments. Twelve circular cells were prepared in a block of sandstone, to each of which a plate of glass was fitted. Toads were then placed in these cells and buried beneath three feet of earth, where they were left for over a year. Every toad shut up in sandstone died ; but the greater number of those in the porous limestone were still alive, though greatly emaciated : these were again shut up ; the end of the second year every toad had died. I also enclosed four toads in holes cut in the trunk of an apple-tree, and closed the holes with a plug of wood ; all these toads were found dead at the end of a year. It seems from these experiments to follow that toads cannot live a year totally excluded from atmospheric air, or two years entirely excluded from food. Admitting that toads are found in cavities of stone and wood, we may account for it by supposing that the toad seeks a cavity while in the tadpole state, and feeds on insects which, like itself, seek shelter within such cavities. It then becomes too large to leave the hole ; but there is always some small crack by which air and food can come in to support life. This tiny aperture is very likely to be overlooked by workmen, who are the only people whose work on stone or wood leads them to disclose cavities in these substances. No examination is made until the toad is discovered by breaking the mass in which it was contained, and then it is too late to ascertain, without carefully replacing every fragment (and in no case that I have seen reported has this been done), whether or not there was any crevice or hole by which the animal may have entered the cavity."

The following anecdote illustrates Buckland's practical activity and ingenuity in gaining information. When the turrets of " Tom Tower " of Christ Church, Oxford, were undergoing repair during the long vacation, he had reason to suspect that all was not right. It was impossible for the Canon to ascend by the slender scaffolding to these turrets ; so, from the windows of his house at Christ Church, he bethought him of watching the masons through an excellent telescope, which he used to examine distant geological sections, etc. At last the unsuspecting mason, working, as he thought, far above the ken of man, put in a faulty bit of stone. Buckland, on the watch below, detected him through the telescope, and almost frightened the man out of his wits when, coming into the quadrangle, he admonished him to bring down directly " that bad bit of stone he had just built into the turret."

The year in which he obtained his canonry was also the year in which he married. His wife was Mary Morland, the eldest daughter of Mr. Benjamin Morland, of Sheepstead House, near Abingdon. The marriage took place on December 31st, 1825. In a letter to the Rev. W Vernon Harcourt, Buckland thus announces the approaching event : " I'm speedily about to follow your example in entering into the holy estate, and propose early in the beginning of the year to set off for Italy and Sicily on a tour of nine or ten months ; if you have any commissions in those regions, pray send them me."

Mary Morland, whose mother died when she was only an infant, was the eldest of a large family of half-brothers and sisters. The greater part of her childhood was spent at Oxford, where she resided with the famous physician Sir Christopher Pegge, whose childless wife took great delight in the lovable and intelligent child. In the University City, and, perhaps, through her acquaintance with the learned Professor of Mineralogy, she acquired that love of natural science which was such a joy to her through all her life. Within a few hours of her death she was working at the microscope, ever looking expectantly for a clearer light in the next world to be shed on the wonders learnt here. Sir R. Murchison, writing of the happy union between Buckland and his wife, calls Mrs. Buckland " a truly excellent and intellectual woman, who, aiding her husband in several of his most difficult researches, has laboured well in her vocation to render her children worthy of their father's name."

Miss Caroline Fox, in her journal for October 8th, 1839, records the following story, which may have some foundation in fact :—

" Davies Gilbert tells us that Dr. Buckland was once travelling somewhere in Dorsetshire, and reading a new and weighty book of Cuvier's which he had just received from the publisher ; a lady was also in the coach, and amongst her books was this identical one, which Cuvier had sent her. They got into conversation, the drift of which was so peculiar that Dr. Buckland at last exclaimed, ' You must be Miss Morland, to whom I am about to deliver a letter of introduction.' He was right, and she soon became Mrs. Buckland. She is an admirable fossil geologist, and makes models in leather of some of the rare discoveries."

The wedding was celebrated at Marcham Church,[1] near

---

[1] Dr. Buckland discovered a curious stone in Marcham Church, which

Abingdon, the Rev. G. Wells, Rector of Boseford, performing
the ceremony. The ringing of the wedding bells was
the signal for the old man-servant to fire off a gun at
Sheepstead House, where Mr. Morland lived. The reason
of this salute was not made known to the family till some
days afterwards. Mrs. Buckland, before her marriage, had
a beautiful white Spanish donkey, which she used to drive
about for miles in a little chaise, in search of freshwater
and land shells, of which she made a very fine collection.
The animal, which was a great pet, had grown very old,
and the servant had been told to make away with it,
without letting the family know when its end came. He
was much pleased with his ingenuity, therefore, in firing
this *feu-de-joie* and at the same time despatching the
donkey.

The wedding tour, which was spent upon the Continent
and lasted nearly a year, is described with much minuteness
and vivacity by Mrs. Buckland in a journal which from her
girlhood she had been in the habit of writing. Naturally
enough her early entries have a geological flavour, and the
scenery and associations of the spots visited are perhaps
less carefully described than the character of the rocks.
The diary, under date of February 25th, 1826, describes
the visit of the newly married couple to Paris. They
called on Humboldt and Arago, and had much scientific

---

he mentions in one of his Ashmolean papers. This stone is said to be
an Anglo-Saxon dedication stone ; the word AELEGY AELEGY-HOLY was
repeated twice on it, and as the church is dedicated to All Saints, it is
suggested that the word might have occurred again in the third line and
have been broken off.

talk with the latter, who showed them his instruments at the Observatory and described his experiments and discoveries.

" Arago," writes Mrs. Buckland, " is the most Englishlike Frenchman I ever saw ; the most unpretending person possible in his manner, and the most intelligent in his conversation.

" From the top of the Observatory," she continues, " we saw Paris in its full extent, built within the basin called by its name, and which is surrounded by low hills, of which Montmartre is the highest. Compared with London, it looks very small, and the absence of smoke gives a coldness so peculiar that it looks like a city of the dead."

Before her marriage Mrs. Buckland had been in correspondence with Cuvier, and had made drawings for his works. She and her husband had now the pleasure of receiving his hospitality and of spending a morning with him in the Jardin des Plantes. The famous naturalist welcomed them with much kindness, and at his house they met Cordier, at that time the most distinguished geologist in France. " The Cuvier's parties," writes the young wife, " are by no means brilliant; he is very taciturn, and so cautious that he never utters an opinion in company ; but though so cold in appearance, he is very friendly in his conduct."

From Paris the travellers journeyed southwards.

" At Vaucluse," says the diary, " when we turned into the steep ravine through which the Sorques flows, we were agreeably surprised by the picturesque forms of the rocks, which are nearly destitute of vegetation, and have nothing but their form and their dazzling whiteness to recommend them to the traveller in search of the picturesque. But to me they possessed other attractions, for I found several varieties of land shells which were new to me, and in the fountain an abundance of nerites and patellæ. The spring or fountain of Vaucluse, from which the Sorques takes its source, is a subterranean river (a phenomenon not uncommon in the Alps), which issues immediately from the bottom of a cliff whose height must be at least eight hundred feet from the basin.

"The quarries near the Pont du Gard and the aqueduct itself are a very coarse *calcaire grossier*, as like as possible to our Norfolk crag, but the mountains which support the Pont on either side are a compact Jura limestone. The great aqueduct, part of which is the Pont du Gard, is still traced to Nismes, and is chiefly underground. It has been much destroyed by persons using the materials for building In the part forming the aqueduct over the Gard, a crust of stalactite has been formed more than a foot thick. We came to Massa, which is the capital of the little duchy of Massa, now in the possession of the Archduchess, mother to the Duke of Modena, who will inherit it. Buonaparte united this with Lucca, and gave it to his sister, Madame Eliza, who was a wise and good princess and did much for her little kingdom. To her the people owe an excellent road to the town of Carrara, which is situated in a fertile valley at the foot of the ridge of marble mountains so famous in all ages. The marble is associated with argillaceous and mica slate, but no granite occurs ; you see the schist passing into marble, by degrees, till the whole is a mass of limestone, which changes its colour according as it possesses iron. Thus fine pure white passes into every shade of bluish grey. We walked to the quarry from whence the statuary marble is procured, and were surprised to find so little of fine quality ; indeed it rarely occurs, and is, consequently, very expensive, being a louis d'or per square foot. A characteristic of the best is that it is highly sonorous."

An incident, that occurred at Palermo on the wedding

tour, was in 1851 still remembered and told by the Consul to illustrate Dr. Buckland's acuteness in the observation of bones.

The Patron Saint of Palermo is Rosalia, daughter of a distinguished nobleman, and born about 1130 A.D. At the age of twelve, Rosalia fled from her father's house to the neighbouring mountains, and passed her whole time in devotion and penance. At length she retired to a cavern on Monte Pelegrino, where she died; but no one knew of this retreat. During the plague of 1624, when all efforts to stay its ravages proved ineffectual, Rosalia appeared in a dream to a citizen at Palermo, and revealed to him where her bones lay unburied. The bones were reverently gathered up and placed in charge of the Archbishop; yet still the pestilence raged. At last a certain man named Vincenza Bonelli, as he wandered on the mountain, encountered a beautiful damsel, who told him she was Rosalia and showed him her grotto. Bonelli, plucking up courage, asked her why she abandoned the city to such cruel ravages. She answered, " It is the will of Heaven; but I am now sent to declare that, as soon as my bones are carried in procession through the city, the plague will be stayed." Bonelli told his confessor of this meeting, and, in obedience to the saint's commands, the relics were carried in procession through the city, and the plague ceased.

The grotto, thus miraculously revealed, was consecrated, a magnificent shrine erected, and a statue to the saint placed there. The bones lay exposed to view behind a *grille*, and the faithful flocked to the shrine.

When Dr. Buckland was at Palermo on his wedding tour in 1826, he, as all strangers did, visited the shrine, and with his keen eyes saw in a moment that the bones never belonged to Rosalia. "Those are the bones of a goat," he said, "not of a woman!" Of course the priests were greatly scandalised, and declared that the saint would not permit him to see what only the faithful could discern. From that time, however, the bones were enclosed in a casket, and neither faithful nor heretics were any longer permitted to scan the sacred relics too closely.

It was on this tour that the Professor recognised the comparatively late geological date of the great upward movement of the Alps, and declared some of the highly inclined rocks to be contemporaneous with our lias and oolite.

On their return journey from Italy, Buckland and his wife visited the cavern of Lunel, near Montpellier, which yielded to the Professor's strong arms and capacious bags many valuable spoils, which were deposited in the Oxford Museum.

His account of this visit, which was made for the purpose of instituting a comparison between Lunel and the caves of England, is extracted from a volume of the Proceedings of the Geological Society.[1]

"The result of the examination has established nearly a perfect identity in the animal and mineral contents of the caverns, as well as in the history of their introductions. In working a free-stone quarry of *calcaire grossier*, the side of the present cavern was accidentally laid open ; it is a long

---

[1] Proceedings of the Geological Society, 1826—1833.

rectilinear vault of nearly a hundred yards in length, and of from ten to twelve feet in width and height. The floor is covered with a thick bed of diluvial mud and pebbles, occasionally reaching almost to the roof, and composed at one extremity chiefly of mud, whilst at the other end pebbles predominate. Stalactite and stalagmite are of rare occurrence in the cavern of Lunel; hence neither its bones nor earthy contents are cemented into a breccia. On examining the bones collected in the cavern by M. Marcel de Serres and his associate M. Cristol, Dr. Buckland found many of them to bear the marks of gnawing by the teeth of ossivorous animals; he also discovered in the cave an extraordinary abundance of balls of *album græcum* in the highest state of preservation. Both these circumstances, so important to establish the fact of the cave of Lunel having been inhabited, like that of Kirkdale, as a den of hyenas, had been overlooked by the gentlemen above mentioned. The more scanty occurrence of stalactite and the greater supply of *album græcum* in this cavern are referred to one and the same cause—viz., the introduction of less rain water by infiltration into this cave, than into that of Kirkdale; in the latter case a large proportion of the remains of the hyenas appear to have been trodden upon and crushed at the bottom of a wet and narrow cave, whilst at Lunel they have been preserved in consequence of the greater size and dryness of the chamber in which they were deposited. The animal remains contained in this cavern differ little from those of Kirkdale; the most remarkable addition is that of the beaver and of the badger, together with the smaller striped, or Abyssinian, hyena."

In October of the same year, while still on his wedding tour through the east of France, he visited the Grotte d'Ozelles, or Quingey, on the banks of the Doubs, five leagues below Besançon. He described the Grotto in a French article, from which the following extract is taken :—

" La grotte d'Ozelles si célèbre par son étendue et par

7

la quantité extraordinaire et la beauté de ses stalactites—
je résolus de la visiter dans le but de m'assurer si elle
ne présentait pas quelque phénomène semblable à ceux des
cavernes à ossemens d'Allemagne et d'Angleterre. . . . A
l'endroit où est située la caverne, une haute colline,
composée de la variété compacte du calcaire Jurassique,
forme la rive gauche du Doubs, et s'élève sous un angle
trop aigu pour permettre la culture à la charrue. L'on
entre par une ouverture de la grandeur d'une porte de
chambre, à peu près de six pieds de haut et de trois ou
quatre pieds de large. Cette ouverture est à environ
cinquante pieds au-dessus du niveau de la rivière. . . .

"Les colonnes et les masses de stalactites qui remplissent
une grande partie de l'étendue de la grotte, excèdent de
beaucoup en nombre, et égalent en beauté celles de la
célèbre caverne de l'île de Sky ou d'aucune autre caverne
que j'aie jamais vue, et l'imagination des visiteurs qui
m'ont précédé, s'est plue à leur y faire trouver toutes les
espèces de ressemblance qu'elle pouvait leur fournir entre
ces stalactites et des animaux, des végétaux ou des
morceaux d'architecture ; mais personne avant moi n'avait
songé à chercher des ossemens sous la croûte de stalagmite
qui s'est accumulée au pied de ces stalactites, et a
formé sur le sol un large tapis ou pavé de differentes
épaisseurs. . . . Ce ne fut pas sans peine que je parvins à
persuader à mes guides de m'aider à rompre cette surface,
jusqu'alors laissé intacte, afin d'y rechercher des restes
d'animaux et de détritus diluvien que, d'après l'analogie
qui existe entre cette caverne et d'autres je m'attendais à
trouver dessous ; leur surprise fut très grande de voir ma
prédiction se vérifier à l'égard de l'existence d'un lit de
limon mêlé de fragmens de pierres et de cailloux roulés,
au-dessous de ce qu'ils considéraient comme le pavé solide
et impénétrable du souterrain, et leur étonnement augmenta
encore, en trouvant à chacune les quatres places que je
choisis pour mon expérience, ce détritus accumulé à une
profondeur que nous ne pûmes percer avec une barre de fer
de trois pieds de longueur, et de plus entremêlé d'une grande
quantité de dents et d'os fossiles. . . . Ainsi, . . . l'absence

de toute marque de dents sur les plus grands os, tendent
à improuver l'action destructrice des Hyènes dans cette
grotte, et à montrer que les Ours en étaient les principaux
habitans. Ces os lorsqu'ils sont secs happent fortement
à la langue comme tous les ossemens anté-diluviens des
autres cavernes. Vers le centre de cette file de grottes, l'on
arrive dans la plus spacieuse de toutes appelée la salle à
danser, parceque sa grandeur et l'égalité du sol l'on fait
choisir pour l'endroit où se rafraichissent et où dansent les
personnes qui viennent voir les singulières beautés de ce
lieu. Cette chambre a, dit-on, plus de cent pieds de long,
et dans quelques endroits cinquante de large.

" Je désire mentionner un indice auquel j'ai déjà fait
allusion, et que j'ai trouvé très utile pour faire distinguer
les os anté-diluviens, que l'on rencontre dans les pentes
et les crevasses, de ceux des animaux récens qui, dans
les temps modernes, se sont introduits dans les mêmes
ouvertures, et par accident ont été mis en contact avec
des restes anciens d'espèces éteintes. C'est la propriété
de happer à la langue, lorsqu'on les y applique tandis
qu'ils sont secs, propriété qui apparemment dérive de la
perte qu'ils ont éprouvée de gélatine animale, sans qu'elle
ait été remplacée par aucune matière minérale.

" La propriété de happer n'appartient que très-rarement
aux os de toutes espèces d'alluvion ou de tourbière, et
n'existe pas non plus dans les ossemens humaines que j'ai
examinés, qui venaient des tombeaux romains d'Angleterre
et des tombes druides des anciens Bretons, ni dans aucun
de ceux que j'ai découverts dans les cavernes de Paviland,
Barrington et Wokey Hole, et que j'ai décrits dans mon
ouvrage intitulé : ' *Reliquiæ Diluvianæ.*' "

The love of natural history in all its branches made
Buckland's home at Christ Church a scene of animal life
not a little strange to the reverend and learned persons
who visited its owner. The house, which was destined to
be the Professor's home for twenty happy and active years,

is thus described by Mr. Thomas I. Sopwith, the famous
mining engineer, who visited it in the "thirties":—

"Dr. Buckland's house is one of those venerable fabrics
which form the principal quadrangle of Christ Church.
As soon as the old-fashioned door is opened, abundant
evidence is presented that the residence is that of a zealous
disciple of Geology. A wide and spacious staircase has
its floor and even part of its steps covered with ammonites,
fossil trees and bones, and various other geological fragments,
and in the several apartments piles upon piles of books
and papers are spread upon tables, chairs, sofas, book-
stands, and no small portion on the floor itself."

Writing when he was again the Canon's guest, and had
the pleasure of meeting Mr. Ruskin, Mr. Sopwith says :—

"Dr. Buckland's house is truly characteristic as the
residence of a geologist and scholar.   In the breakfast-
room was a series of piles of books, boxes, and papers ; in
short, such a combination of book-stands, chairs, sideboards,
boxes, all blended together in one mass of confusion,
which I was informed had not been invaded by the dust-
cloth for the last five years. The drawing-room at Dr.
Buckland's had its share of variety, and the great interest
of a tolerable deal of confusion through which a person
might range a whole day and find some new index every
moment pointing to weeks and months and years of
occupation. One of the round tables is formed entirely of
coprolites. Another presents on its highly polished surface all the variety of lava, etc., found at Mount Etna."

Mr. Sopwith adds that "the most interesting part of
this interesting mansion is the domestic comfort which so
eminently prevails." The writer judged truly that the
Christ Church home was a happy one. Buckland and his
wife had a large family,—nine altogether, four of whom

were buried in the vault in Christ Church Cathedral. The surviving children were all blessed with excellent health, good tempers, and loving dispositions. If they never quarrelled, the reason must have been that they never had idle playtime. There was always something to do,— their animals to feed, or their gardens to tend, or, if a wet day came, they all adjourned to the dining-room and sat round the big table helping Mrs. Buckland to cut and paste cardboard into strong neat little trays for specimens, while one of the party read aloud, generally from a book of travel or Arctic voyage. If the book was not illustrated —and illustrated books were rare fifty years ago—Mrs. Buckland would be sure to have found some old pictures or illustrations of some sort to show them on the subject. Like their father, she never taught her children without a picture or a piece of paper and charcoal at hand. To give zest to their Bible readings they had some quaint engravings in Mrs. Trimmer's two little square books. Anthony Trollope once told one of the children, when, years later, they were talking together over the days of their youth, that his mother used the same little books for him, and that he "loved them."

On one point only Dr. Buckland was a strict father. He never allowed his children to be unemployed. Those who were too young to work, folded up old letters, kept ready to be made into spills for the lighting of their father's Winchester Taper, which he always used to read by. When postage stamps first came into use, it was the children's work to cut them up and stick them on the envelopes. Pennies were earned for doing these little tasks neatly and quickly, and the only pocket money which the children had, was earned by the quantity and quality of the work that they did. They were taught Dr. Watts' hymns, and their mother, if she ever found them unemployed, would make them repeat the lines : " Satan finds some mischief still for idle hands to do." Their father attributed almost every disaster to laziness, " which was," he said, " the root of all evil."

The family life of the Bucklands is described so vividly in the " Life of Frank Buckland," that no apology is needed for extracting a passage from that volume :—

" In his early home at Christ Church, besides the stuffed creatures which shared the hall with the rocking-horse, there were cages full of snakes, and of green frogs, in the dining-room, where the sideboard groaned under successive layers of fossils, and the candles stood on ichthyosauri's vertebræ. Guinea-pigs were often running over the table ; and occasionally the pony, having trotted down the steps from the garden, would push open the dining-room door, and career round the table, with three laughing children on his back, and then, marching through the front door, and down the steps, would continue his course round Tom Quad. In the stable yard and large wood-house were the fox, rabbits, guinea-pigs and ferrets, hawks and owls, the magpie and jackdaw, besides dogs, cats, and poultry, and in the garden was the tortoise (on whose back the children would stand to try its strength), and toads immured in various pots, to test the truth of their supposed life in rock-cells. There were also visits to the Clarendon, where Dr. Buckland was forming the nucleus of the present Geological Museum of Oxford, and to the Ashmolean Museum, then under the wise and genial care of the brothers, John and Philip Duncan, where the children might ride the stuffed zebra, and knew all the animals as friends, if not yet as relations.

"In summer afternoons, after the early three o'clock
dinner, Dr. Buckland would drive out Mrs. Buckland and
their children in a carriage, known as the bird's-nest, to
Bagley Wood, to hunt for moles and nests, or to Port
Meadow to gather yellow iris and water-lilies, and fish
for minnows, and often to set free a bright-hued king-

PROFESSOR AND MRS. BUCKLAND AND FRANK.

fisher (they were plentiful in those days) which he had
redeemed from some mischievous urchin with a sixpence.
Or another day to Shotover, to dig in the quarries for
oysters and gryphites ; or again to Iffley, to gather snakes'
heads (*Fritillaries*). Both father and mother were devotedly
fond of flowers, and their horse stopped automatically at
every nursery garden, as at every quarry. Some of the
graver dons were perhaps a little scandalised by such

vagrant proceedings, but how much happiness and wisdom were gathered in these excursions!"[1]

The Rev. George Gaisford used sometimes to tell the story of his watching with Frank at the window to see the Dean's (Gaisford) carriage as it passed round the corner of Tom Quad. The moment it was out of sight, he turned to Frank and cried, "Now then, Frank, let's put the crocodile into Mercury" (the pond in the middle of Christ Church Quadrangle, so called from a little stone statue of Mercury in the centre, used as a fountain).

Mr. Ruskin writes in "Præterita" :—

"At the corner of the great Quadrangle of Christ Church lived Dr. Buckland, always ready to help me,—or, a greater favour still, to be helped by me, in diagram drawing for his lectures. My picture of the granite veins in Trewavas Head, with a cutter weathering the point in a squall, in the style of Copley Fielding, still, I believe, forms part of the resources of the geological department. . . . At his breakfast-table I met the leading scientific men of the day, from Herschel downwards, and often intelligent and courteous foreigners. . . . Every one was at ease and amused at that breakfast-table,—the menu and science of it usually in themselves interesting. I have always regretted a day of unlucky engagement on which I missed a delicate toast of mice ; and remembered with delight being waited upon one hot summer morning by two graceful and polite little Carolina lizards, who kept off the flies."

"Your father the Dean," Lord Playfair writes to Mrs. Gordon, "was a born experimentalist, and I recollect various queer dishes which he had at his table. The hedgehog was a successful experiment, and both Liebig and I thought it good and tender. On another occasion

---

[1] "Life of Frank Buckland," pp. 8, 9.

I recollect a dish of crocodile, which was an utter failure.
The Dean's experiment in quaint gastronomy used to
remind me of the dinner on garden snails at which Black,
Hutton, and Playfair determined to get over their natural
prejudice; but though the three philosophers took one
mouthful, they could not be persuaded to swallow it, and
rejected the morsel with strong language. The crocodile
at your father's table had a similar fate."

On the opposite side of the Christ Church Quadrangle
lived Dr. Pusey, who was a most kind friend and neighbour
to both Dr. and Mrs. Buckland, and his spiritual minis-
trations afforded much comfort to Mrs. Buckland at the
time of the death of her son Adam when only nine years
old. Adam, also called Conybeare Sedgwick after his
godfathers Dr. Conybeare, Dean of Llandaff, and Professor
Sedgwick, was buried in Christ Church Cathedral, in a
vault where already lay two elder children—by name
Willie and Eva.

Buckland was a kind and affectionate father, and always
liked to have his children about him. The return from
his frequent journeys was awaited by them with eager
expectation, for from the famous blue bag would be turned
out for them on the dining-room floor some strange (in
those days) foreign fruit, such as a bundle of bananas,
or a cocoanut in its big outside shell, or a "forbidden
fruit" (lime), which the little ones fondly imagined might
have grown in the Garden of Eden. On one occasion,
in addition to the blue bag, a large mysterious bundle
was brought in, wrapped in a travelling rug. The children
were told that it was a "wild beast" of some sort, that
it would not hurt them, and that whoever guessed what

it was would be rewarded with a penny. The wild beast proved to be the carcases of a bear, which had been seen hanging up outside a barber's shop as an advertisement for the celebrated Bear's Grease—a pomatum for the hair, then much in vogue. The beast had been prepared just like a sheep at the butcher's, only that the skin had been left on the head and hind-legs to show that it was the veritable animal. A luncheon party was invited to partake of joints of bear, and the fat from the inside was given to the nurse to make into pomatum for the family use.

The young people were always presented to the numerous learned foreigners and illustrious travellers who came to Oxford to see the Professor's world-famed collection of fossils and bones at the Clarendon; and at dessert in the evening they were told, shortly and graphically, what these great men were famous for. They heard that Agassiz came from Switzerland, and how he once lived in a little hut on a glacier in order to watch the frozen river slowly move down the valley between the snow mountains about one inch a year; that Liebig was a great German chemist, Sir John Franklin the famous Arctic voyager, Warburton an African traveller, and so on. Occasionally one or two ill-clad foreigners with very large appetites would be entertained with boundless hospitality and courtesy. At such times the children would listen with curiosity to the rapid talk in a strange language, and watch the lively gesticulations, and wonder how, at the same time, the speakers could manage to empty plate after plate of food.

Mrs. Buckland took great interest both in the spiritual

and bodily welfare of a settlement of Jews living in St. Ebbe's parish, a very poor part of Oxford. When several families were once burnt out of house and home, she greatly befriended them. One man she set up as a pedlar ; and for many years he travelled about the country with a mahogany box strapped at his back. Twenty years after he had thus commenced business, the family were chiefly living at Islip, during the illness of the Dean. The pedlar was a regular caller at the Rectory, where he would display his wares—silver thimbles, trinkets, and brooches tastefully arranged in trays on pink wadding— and generally sold some silver thimbles, which were bought as gifts to the girls who were the best darners and menders in the village school. These poor Jews, soon after the fire had destroyed all their goods, came to Mrs. Buckland to borrow a glass goblet, which was required for a wedding that was about to take place. Mrs. Buckland, who lent them a handsome cut glass one, was invited to the wedding. She took with her one of her children, a little fellow of five or six. In the middle of the ceremony a glass was smashed ; the child called out at the top of his voice, greatly to his mother's consternation, " Oh, Mamma, there's your best glass broken ! " It is needless to say that a substitute had been provided to be smashed, and that the lent goblet was returned safely. Regularly as the Feast of the Passover came round, half a dozen of the large thin wafer-biscuits, about twelve inches across,—the " Passover Bread,"—were sent as a present to Mrs. Buckland, in token of respect and gratitude from the Jewish community.

Buckland was very fond of his garden, and brought plants from all parts of the Continent to place in it, often with the view of acclimatising them. A fig tree, which is still in existence at Christ Church, was brought from Aleppo by Dr. Pocock. The Professor was especially fond of sweet-scented flowers, and it was the children's business to provide him always with a Sunday button-hole. The first violet and cowslip were actively searched for, to be succeeded by woodbine, cabbage-rose, southern-wood (old man), jessamine, and clove-pink ; and then, when all the flowers were gone and autumn had set in, a sprig of lemon verbena picked from a greenhouse plant brought from Sicily.

Dr. and Mrs. Buckland trained their children to take an interest in the conversation going on at meals, and always, when they came down to dessert in the evening, their father would have some anecdote or curious fact to tell them, and they on their part were expected to have something of interest to tell him—or some question to ask. This was no great difficulty, for the children were never taken for a dull "constitutional" walk, but were always sent on some special errand. Sometimes, for instance,

---

Edward Pocock, Chaplain to English merchants, Aleppo, 1620. The Laudian Professorship was founded in his honour by Archbishop Laud, 1632. He was Professor of Hebrew and Canon of Christ Church, and was sent at Laud's charges to purchase and collect Arabic manuscripts now in the Bodleian Library. Professor Margoliouth, who is at present indexing these manuscripts, adds that Pocock was highly distinguished as a theologian and Orientalist; and that in the opinion of Hallam he did more than any other one man to familiarise Europeans with the East.

their errand would be to take some "alicampane"—an
old-fashioned herbal remedy made up with sugar in pink
and white squares, bought from some old Meg Merrilies
in the market—to a barge-man with a bad cough, who,
with the aid of his family, was unlading the barge as it
lay under the shadow of the fine old Norman Keep from
whose postern gate, as they were told, the Empress
Maude escaped in a white sheet over the frozen river to
Abingdon.

This old-world corner of Oxford, with the high earth
mound adjoining, and the Gaol, or Castle, as it was then
called, was always full of mysterious interest to the little
people. There were no railways then, and several gaily-
painted barges were often to be seen moored along the
Canal Wharf, supplying the city with coals, salt, or pottery.
However grimy their cargo might be, the owners contrived
to keep fresh and bright the gay lines of colour on the sides
of the little cabin at the end of the long black hull. Dr.
Buckland, or occasionally a good-natured bargee, would lift
the children into the empty barge and allow them to peep
into the snug little abode, reeking with the savoury smell
which issued from a black iron pot on its small hob, while
from the tiny low chimney-pipe curled the prettiest
possible wreath of blue-grey smoke.

Never was a word of gossip or evil speaking permitted ;
the good clever mother would always say, " My dears,
educated people always talk of things ; it is only in the
servants' hall that people talk gossip." Thus the family
were trained from childhood to live in charity with all
men.

One summer day, on the Duke of Cumberland visiting Oxford, Dr. Buckland had undertaken to lionise him and a number of gentlemen. The children, returning from their favourite walk by the house-boats in Christ Church meadows, came across the distinguished party. "Come here, children," said Dr. Buckland, "and make your curtseys to the Royal Duke." The kind old gentleman patted one on the head, and said, "How old are you, my little maid?" "Please, sir, I am ten." "How can you say so?" exclaimed the more truthful younger sister; "you are only nine and a half." At which the Duke laughed heartily, and said, "Little lassie, you'll not be so anxious to make yourself older when you have lived more years."

Dean Gaisford was very fond of these two little girls. Whenever Dr. and Mrs. Buckland dined at the Deanery, the Sedan chair—a most convenient conveyance for collegiate buildings—which had carried Mrs. Buckland to the six o'clock dinner, was sent for the children, who were carried safely into the large dining-room, and took their places on either side of the kindly old Dean, whose pleasure it was to keep them well supplied with dessert.

The Sundays at Christ Church were the children's red-letter days. Buckland always went to the morning service in the Cathedral, and to the University sermon at St. Mary's. It was their never-failing delight to watch the procession of the Vice-Chancellor, preceded by the beadle and college dignitaries, students, and graduates in their robes, wending its way to the Church of St Mary in the High Street, whose spire is one of the glories of the city, and entering by the porch which has in a niche over the door an image of the Blessed Virgin. In 1640 Archbishop Laud was charged with many offences. He had repaired crucifixes ; he had allowed " the scandalous image " to be set up in the porch of St. Mary's ; and Alderman Nixon, the Puritan grocer, had, so he declared, seen a man bowing to the scandalous image. Alderman Nixon's picture, with that of his wife, is to be seen in the fine old council chamber of the Guildhall. This lady's portrait, it may be added, is the only likeness of a woman admitted among the interesting collection of Aldermen, with the exception of a fine full-length portrait of Queen Anne. Mrs. Buckland took the younger members of the family to the simple morning service at St. Ebbe's Church, of which the Rev. F. Waldegrave, an excellent evangelical preacher, was in charge. Mrs. Buckland was an assiduous worker in Mr. Waldegrave's poor parish.

After the early dinner came the treat of the week—a walk with their father in Christ Church meadows, or, if the floods were out, up Headington Hill. No plant, tree, or stone escaped observation, and special notice was taken of the dates of the reappearance of palm blossom, or the first return of daisies, and other spring delights. The family never missed evensong in the Cathedral. The seat allotted to the Canon's ladies was like a very long saloon railway carriage, with a seat running along one side of it. As this pew had only occasional oval openings in the heavy wood-work to admit the light and air, its dreariness and stuffiness may be imagined.

The yearly assizes were a great annual function in

Oxford, and were very awe-inspiring, solemn events. The week began with an assize sermon at St. Mary's, which the Judge attended in state. Dr. Buckland used to take his children to see the stately procession of the Judge, and other dignitaries of the law, as it traversed the whole length of Oxford from the Judge's house in St. Giles' to the Town Hall in St. Aldates'. The Judge, attired in wig and robes and seated in his ponderous coach, was driven by a fat coachman, whose mighty weight seemed to have caused a depression on the box-seat, covered with hammer cloth fringed round with gorgeous coloured tassels ; two or three footmen stood behind the coach, long wands in hand. At the entrance to the Town Hall, the beadles were collected to keep order, for there were of course at that time no policemen. These pompous individuals wore a quaint dress and cocked hats, long frock-coats lined with red with brass buttons, and carried in their hands stout wands of office.

The prisoners were brought from the old castle to the Town Hall to stand their trial. The new gaol with its adjacent judgment hall or law court was not then built. The Rev. Gilbert Heathcote, Sub-Warden of Winchester, tells a curious tale which shows the inconvenience arising from the want of space in the Town Hall buildings. When he was an undergraduate, he attended Dr. Kidd's lectures. The Regius Professor of Medicine had a male and female skeleton suspended from the ceiling, on either side his lecture table, which he could pull up and down as required. The male skeleton was of almost gigantic stature, and was that of a man who was tried for murder and convicted in the Town Hall. The bodies of the criminals in those days were handed over for anatomical purposes to the Professor of Medicine. This big culprit, finding the case was going against him through the evidence of a witness, stretched out his long arm over the witness box, and with one mighty blow felled the unfortunate man to the ground and killed him on the spot, so that one murder begat another.

On another occasion, the Judge had passed sentence of imprisonment upon a woman. As she was leaving the dock, she took off one of her shoes (boots were not worn in those days), and threw it with such good aim and with so good a will at the Judge, that he was not a little discomfited—in fact, his Lordship was nearly sent reeling from the Bench.

After the death of the little boy Adam, the family went by coach for change of air to Lyme. The shore in this neighbourhood is a vast charnel-house of fossil bones of the monsters that must have at one time lived, preyed on one another, and ultimately died, at or near this very spot. On the lias beds of this happy hunting ground of geologists, Dr. Buckland took the children fossilising, and made them acquainted with the local celebrity Mary Anning, who, from the early age of ten, gained her livelihood and supported a widowed mother by collecting specimens on the beach. It was in 1811 that she made her first great discovery of the ichthyosaurus, which, with the vertebræ of a fish, partook partly of the character of the crocodile, but differed materially from any existing reptile of the lizard kind. At Lyme also lived Sir Henry

8

de la Bêche, who had ample leisure and opportunity to picture to himself the shape and habits of the former dwellers on this sea-girt coast, as, by the daily action of the tides, vertebræ (called there verteberries) and portions of shells and skeletons were exposed to view, or washed up from their bed of soft blue lias. Here the remains of extinct monsters were picked up or disinterred as "curiosities" by Mary Anning, described for the first time by Buckland, and restored to life by the clever pencil of Sir H. de la Bêche. Some of the largest of the ichthyosauri were over thirty feet long, the jaw sometimes exceeding six feet. They were aquatic carnivorous animals, but breathing air.

"When we see," says Dr. Buckland, "the body of an ichthyosaurus still containing the food it had eaten just before its death, and its ribs still surrounding the remains of feeding that were swallowed ten thousand or more than ten thousand times ten thousand years ago, all these vast intervals seem annihilated, come together, disappear, and we are almost brought into as immediate contact with events of immeasurably distant periods as with the affairs of yesterday."

Miss Anning received the sum of twenty-three pounds from the British Museum for this specimen. Later on she discovered the plesiosaurus, another of these extinct monsters. It must not, however, be supposed that these immense fossils, which we see so admirably arranged in the Reptilian Gallery of the British Museum of Natural History, were extracted from the rock in which they had been embedded for ages without considerable trouble and perseverance; often the remains were found in a fragmentary condition, and the greatest judgment and care were required in arranging the disconnected parts.[1] Miss Anning kept a little curiosity shop at Lyme, which is admirably described in the King of Saxony's account of his journey through England and Scotland in 1844 :—

"We had alighted from the carriage, and were proceeding along on foot, when we fell in with a shop in which the most remarkable petrifactions and fossil remains—the head of an ichthyosaurus, beautiful ammonites, etc.—were exhibited in the window. We entered, and found a little shop and adjoining chamber completely filled with fossil productions of the coast. It is a piece of great good fortune for the collectors when the heavy winter rains loosen and bring down large masses of the projecting coast. When such a fall takes place, the most splendid and rarest fossils are brought to light, and made accessible almost without labour on their part. In the course of the past winter there had been no very favourable slips; the stock of fossils on hand was therefore smaller than usual : still I found in the shop a large slab of blackish clay, in which a perfect ichthyosaurus of at least six feet was embedded. This specimen would have been a great acquisition for many of the cabinets of Natural History on the Continent, and I consider the price demanded— £15 sterling—as very moderate. I was anxious at all events to write down the address, and the woman who kept the shop, for it was a woman who had devoted herself to this scientific pursuit, with a firm hand wrote her name 'Mary Anning' in my pocket-book, and added, as she returned the book into my hands, 'I am well known throughout the whole of Europe.'"

---

[1] It took Miss Anning ten years to extract the entire skeleton of the plesiosaurus from its watery grave in the lias rocks, only accessible at low water. Lately a man has spent two years of patient labour in extracting from its rocky matrix the fossil skeleton of a turtle from the Cape, which is now placed in the British Museum of Natural History.

The tiny old "curiosity shop" close to the beach is still
in existence ; but there are none of the pretty little boxes
of shells or tastefully arranged bunches of seaweed of former
days to be seen now. Foreigners long continued to write
for specimens, little realising that the moving spirit and
indefatigable collector of these old-world treasures had
passed away. Miss Anning's collection was broken up at
her death. The best portion of it passed into the hands
of the Misses Philpot, and is now in the Natural History
Museum, South Kensington ; but a small part of the collection is at Oxford. Mary Anning was born in 1800, and
died in 1847. Buckland succeeded in obtaining an annuity
for her. A stained-glass window was erected to her
memory in Lyme Church, with the following inscription :—

"This window is sacred to the memory of Mary Anning,
of this parish, who died March 9, 1847, and is erected by
the Vicar of Lyme and some of the members of the
Geological Society of London, in commemoration of her
usefulness in furthering the science of geology, as also of
her benevolence of heart and integrity of life."

Before concluding the account of the visit to Lyme Regis,
allusion should be made to the interesting group of fossil
animals discovered in the neighbourhood. An illustration
of these is given, after the drawing of Sir H. de la Bêche,
of which the following is an explanation :—

EXPLANATION OF THE DRAWING BY SIR HENRY DE LA BÊCHE CALLED
"DURIA ANTIQUIOR," OR ANCIENT DORSETSHIRE.

Dr. Buckland always kept a good supply of his old friend's clever
representation of these monster inhabitants of ancient seas, and
frequently after his lectures distributed copies in order to bring to the

ANCIENT DORSETSHIRE.

[Sir H. de la Bèche.
[Face p. 116.

minds of his audience the reality of the subjects on which he had been speaking.

In the centre of the plate, at fig. 1, is seen the mighty Ichthyosaurus, or the Lizard Fish—in form and structure not unlike the marine mammalia of the present day. The Ichthyosaurus was an air-breathing creature, and this is known, firstly on account of there being an entire absence of that peculiar modification of the bones which support the gills in fish; secondly, because there are found true bony nostrils, and not olfactory bags, placed in the skin, unconnected with bone, as in fish; thirdly, the articulation of the ribs to the spine is similar to those in recent air-breathing animals. Ichthyosaurus had fins or paddles at its side, and a long tail, at the end of which, according to Professor Owen's recent discoveries, was a vertical fleshy fin. It could do what no whale or grampus of the present day is capable of accomplishing, viz., could crawl upon the shore, and that most likely at periodical times, as do the seal, walrus, etc. It had an enormous eyeball, which was larger in proportion to the skull than the eye of any other kind of animal; and this eye, having no eyelids, contained delicate humours, which, being liable to injury in a chopping sea, were composed of numerous thin and (probably) flexible bones, which encased the pupil. Owl-like, it probably pursued its prey at dusk of evening, by moonlight, or at early morning. It had a formidable array of teeth, each of which was undermined by the germ of its successor, so that if a violent snap or a too vigorous captured prey broke away the old tooth, the new one would come up in its place. In the engraving it is represented as making good use of these teeth, for it has caught and is about to devour a Plesiosaurus (fig. 3)

This also was a curious whale-like creature, which has aptly been likened to "a turtle threaded through with the body of a snake." This animal was marine-aquatic in its habits; but unlike the Ichthyosaurus, which was a deep-sea animal, it was a shore creature, and lived in the estuaries of brackish water; and there, lurking under the oar-weed and other marine vegetations obtained its prey by darting out its long neck and seizing its prey with its sharp and formidable teeth, as is seen in fig. 4 (and also in the distance), where an unfortunate Pterodactyle has not got out of the way quickly enough, and is suffering for his laziness; while his frightened companions are wheeling about in the air overhead, like frightened seagulls when one of their comrades has been captured or shot.

This Pterodactyle, or wing-fingered saurian, was a monstrous beast, a true saurian, but yet with leather-like wings like a bat; the only living approach to them is the insignificant little Draco-Volans of the isles of the Indian Archipelago.

At fig 5 is seen a fish whose name is Dapedius, so called on account of its " pavement-like " scales ; it has encountered in its peregrinations an Ichthyosaurus, who is making short work of him, and is about to gorge him down into its capacious stomach in the same manner that a jack does a roach or dace. We know that Ichthyosaurus fed upon this fish, because its scales are found in the fossil coprolites.

In the Oxford Museum is the fossil stomach of an Ichthyosaurus that had died shortly after its dinner, as it had not had time to digest entirely the fish it had swallowed. The Ichthyosaurus (as seen in the engraving) did not refuse to eat cuttle-fish, and we know this because the ink of the cuttle-fish is found staining the fossil coprolites.

Other fossil fishes whose remains are found are seen swimming about in company with young Ichthyosauri, all enjoying life, and following the laws of nature which ordained that they should both prey upon one another and be preyed upon themselves.

Sailing along the surface of this sea, upon which no human eye ever rested, may be noticed a fleet of the beautiful Ammonite shells. Their remains are seen at the bottom of the sea, where they would become gradually covered with mud and converted into fossils, a theme for the geologists and for the adornment of our cabinets.

At fig. 6 we see growing in great luxuriance a remarkable form of life—the Pentacrinite, or Stone Lily, so called on account of the pentangular or five-sided shape of its supporting column. It consisted of innumerable calcareous joints, united by a fleshy material; it was, in fact, a " stalk star-fish," which is represented in existing seas by the Comatula, or Feather-star, of our own shores, and by the rare and all but extinct Pentacrinites of the West Indies.

For a full and beautifully illustrated description of the Pentacrinite, as well as of the Ichthyosaurus, Plesiosaurus, Pterodactyle, and other creatures represented in the drawing, I must refer my readers to Dr. Buckland's Bridgewater Treatise.

At fig. 10 is represented the Zamia, or " bird's nest," of the Portland quarry men, together with restorations of vegetation which once flourished in luxuriance, but which is found now only in a fossil state.

At the bottom of the primæval sea are strewed the bones and carcases of its inhabitants, both small and great. Saurians, fishes, molluscs, and shells have all yielded up their remains in obedience to the dictum which pronounces the sentence of death upon everything that has ever been or ever will be animated with the breath of life.

In their unknown graves for thousands of past centuries, converted into hard marble-like rocks, they have lain, and hundreds of skeletons will lie, till time is no more, leaving but a bare record of their former existence engraved in tablets of stone on the shores which once formed the bed of an ancient ocean, now long passed away.

Meanwhile let it be our privilege to read and interpret the history of our planet as it existed when yet young in the starry firmament. Let us compare extinct forms of animal life with their modern living prototypes ; and from the habits and instincts of animals around us, learn, not only the laws which govern them as well as ourselves, the physiological causes which regulate *our* bodies as well as *their* bodies, but also endeavour to learn pleasurable lessons from daily scenes, and to withdraw the veil which frequently obscures the most enchanting scenes of nature from ordinary observation.

Above all, let us join with the inspired writer when he admonishes us : " But ask now the beasts, and they shall teach thee ; and the fowls of the air, and they shall tell thee : or speak to the earth, and it shall teach thee : and the fishes of the sea shall declare unto thee."

CHAPTER V.

THE BRITISH ASSOCIATION AT OXFORD, 1832; THE
    MEGATHERIUM; THE BRITISH ASSOCIATION AT
    BRISTOL, 1836; FOREIGN AND ENGLISH FRIENDS;
    THE QUEEN AT OXFORD, 1841.

1831—1841.

THE year 1831 brought to light the first germ of the
British Association. Mr. (afterwards Sir David)
Brewster proposed that a "craft should be built wherein
the united crew of British science could sail." His
notion found an enthusiastic supporter in the Rev. W.
Vernon Harcourt, a great lover of science, who invited all
Philosophical Societies in Great Britain to meet at York.
Buckland, who was unable to be present, owing to the
death of a child, writes to express his "bitter disappointment" at his enforced absence. He was chosen President
of the next meeting, which was held the following year
at Oxford.

An old geological pupil of his, the Rev. W. Egerton, the
present Rector of Whitchurch, Salop, has kindly placed at
the service of the biographer Buckland's humorous letter
of congratulation to his brother, Sir Philip Egerton,

upon his intended marriage, and visit to Oxford for the meeting of the Association.

"Christ Church,

"*January* 23*rd*, 1832.

"My dear Sir Philip,—Mrs. Buckland begs to unite with me in the offering of our most sincere congratulations to you on the brilliant Discovery announced in your last, of a Jewel of great price which you have resolved to make your own, and to submit to the inspection of the learned, at our proposed scientific meeting in June next.   The only rival specimen I have heard of as likely to be present, and which has the reputation of being the greatest Beauty in the mineral world, is a specimen that will be brought by the Marquis of Northampton, who has joined our Society, and has lately possessed himself of a fossil lizard enclosed in amber more exquisitely beautiful than the fairest of the fossil Saurians, and which your specimen alone I expect to find possessing the power to eclipse.   Your scientific description of that specimen is, I presume, submitted to me as a paper to be read at the meeting, when all who may be present will have opportunity of ascertaining its fidelity by comparison with the original, and of applauding the taste and discretion you will have exhibited by the selection you have made.   I presume our friend Lord Cole will appear in his unenviable state of single blessedness.

" Again repeating our united congratulations, and with most sincere wishes for your happiness, I remain,

" Yours always very sincerely,

" W. Buckland."

To Sedgwick he writes requesting him " by all your love of Professional Unity and the eternal fitness of things to locate yourself in a fraternal habitation within my domicile during the orgies of the next week, beginning the 3rd of June " ; and then goes on to tell him of the arrangements that Mrs. Buckland had made for his comfort, and that of the friends whom he would probably meet. Among these was the Duke of Sussex,[1] who was to be his guest at Christ Church. Sedgwick had rather ridiculed the notion of such a gathering when it was first proposed, and had protested that he would not leave Wales for either York or Oxford. Murchison, however, made him break his resolution in favour of the latter city, and his friend's warm invitation clinched the matter.

It may be said that at Oxford the British Association made her most brilliant *début*. Only thirteen years previously had geology been recognised by the University as a science, when its Professor was appointed, and, after much opposition to the new learning, all Oxford seems to have united in welcoming with boundless hospitality the *savants* of the day. The Vice-Chancellor, Dr. Jones Collier, gave a public breakfast in Exeter College Gardens, and a free supply of refreshments was furnished to the evening meetings in the Clarendon Buildings. The Archbishop of York (Vernon Harcourt) sent a fatted buck from Nuneham Park ; the Duke of Buckingham also supplied venison ; and never before were there witnessed such scientific enthusiasm, goodwill, and friendship among all classes in the old University and Cathedral town.[2]

One burning question which the Committee of the

---

[1] The Duke of Sussex was at this time President of the Royal Society.

[2] "We ascribe the success of the Association exclusively to its *migratory* character. The learned junta, now so gigantic and over-whelming, sprang from a lowly origin. Four years ago a few unpretending individuals, full of zeal for experimental science, met together at York for the formation of a philosophic union, in modest imitation of

Association had to decide was whether or not women were to be admitted to the meetings. " I was most anxious to see you," writes Buckland to Murchison in 1832, " to talk over the proposed meeting of the British Association at Oxford in June. Everybody whom I spoke to on the subject agreed that, if the meeting is to be of scientific utility, ladies ought not to attend the reading of the papers—especially in a place like Oxford—as it would at once turn the thing into a sort of Albemarle-dilettanti-meeting, instead of a serious philosophical union of working men. I did not see Mrs. Somerville ; but her husband decidedly led me to infer that such is her opinion of this matter, and he further fears that she will not come at all." In the end Mrs. Somerville decided not to attend the meeting, for fear that her presence should encourage less capable representatives of her sex to be present. In this respect, as in many others, at Oxford and elsewhere, the lapse of sixty years has made vast alterations. Another change which is not unworthy of notice is in the attitude of the great newspapers towards such gatherings as that of the Association at Oxford. Almost the only, if not absolutely the only, reference to the meeting which occurs in the *Times* is contained in a leading article for June 28th, 1832 : " We have received," says the article,

---

certain ambulatory societies in Germany. These excellent persons, not aware of their own possible importance, formed the most moderate prognostics of success, and were even apprehensive of total failure. Professor Buckland, however, with a generosity most chivalrous, invited the infant body to the hospitable halls of Oxford. Here its numbers doubled, and the celebrity of the place gave celebrity to the institution." —*The Oxford University Magazine, November* 1834.

"some notices from correspondents respecting the character
and proceedings of the present meeting of scientific men
at Oxford," and it goes on to give its reasons for thinking
that such meetings are useless.

The proceedings opened in the Sheldonian Theatre by
the President requesting Mr. Murchison to present to the
"Father of English Geology," Mr. William Smith,[1] the
Wollaston Medal, awarded to him by the Geological
Society. The death of the great Cuvier, which had
recently occurred on May 13th, 1832, called forth from
the President an eloquent and graceful tribute.

"I cannot," he said, "utter the name of Cuvier without
being at once arrested and overwhelmed by recollections
of mortality, melancholy and painful. We have at this
moment to deplore, in common with the whole philosophic
world, the loss of the greatest naturalist, and one of the
greatest philosophers, that have arisen in distant ages to
enlighten and improve mankind. The names of Aristotle
and Pliny and Cuvier will go down together through every
age in which natural science and natural history, in which
philosophic talent and learning, and everything which, next
to religion and morality, give dignity and exaltation to
the character of man, shall be respected on earth. It was
the genius of Cuvier that first established the perfect
method after which every succeeding naturalist will model
his researches. He has shown that the frame and
mechanism of every animal present an uniformity of design
and a simplicity of purpose which prove to demonstration
that every individual, not only of the existing species, but
of those numerous and still more curious species which

---

[1] According to Professor Phillips, in the Life of his uncle, it was at
Dr. Buckland's suggestion that a memorial tablet to William Smith was
placed in All Saints' Church, Northampton, by a subscription among
geologists.

have lived and perished in distant ages, and our knowledge of which is due to discoveries in Geology, was formed and fashioned by the same Almighty Hand. At the age of sixty-three, in the vigour of his mind, he has been called to an early grave. The gratitude of the great nation to whose philosophic fame his genius has added so bright a wreath has already displayed itself by a liberal provision for his family, and has fixed his widow during the remainder of her mortal life in that honoured and well-known mansion in the Jardin des Plantes, which during a quarter of a century has been ever opened in friendly hospitality to every son of science assembled at Paris from every nation under heaven. I fear my feelings of respect and love and gratitude have transported me beyond the limits which the task I have undertaken should impose on me; still I cannot but rejoice in the opportunity which this august assembly affords of inviting you to partake in this great and glorious work, and thus publicly to record your gratitude to that immortal man, whose friendship I have ever counted among the most distinguished honours of my life, and whose genius will be ever venerated so long as science shall be cultivated or virtue venerated upon earth."

Nor was Buckland content with words only. It was, it may be added, mainly owing to his suggestion and active exertions that a considerable sum of money was collected in England, and handed over to M. Cordot, who acted as treasurer of the fund raised in Paris to commemorate the memory of the great philosopher and naturalist.

Among the noticeable events of the week was a lecture delivered by Buckland on the summit of Shotover Hill to a large class of the members, including both veterans in science and ladies. It was at this lecture that, for the first time, attention was drawn to the importance of the application of a knowledge of geology to agricultural improvements. In the course of his remarks the Professor pointed out many defects in the ordinary system of drainage which could be remedied by a knowledge of the structure of the strata, and adverted to the possibility of reclaiming the peat bogs.

Still more remarkable was the interest excited by his lecture upon the megatherium, which was delivered on the last day of the meeting. The occasion was the first on which a fossil monster had been described to an unscientific audience of ladies and gentlemen. The whole address forms an excellent illustration of Buckland's power of imparting interest to the subjects on which he touched. "How true," wrote Sir Richard Owen in 1853 to Mrs. Buckland, "is all that you say in the comparison of the poor Dean's style of communicating knowledge with that of the best of us. His like will never be listened to again! Only those who have heard him can appreciate the loss. It was the most genial inspiration ever vouchsafed to a teacher of the Creator's doings of old."

Though the megatherium does not figure in the sketch given on p. 127, the picture affords an amusing comment on the enthusiasm of the lecturer, whose personality possessed that marked originality and individuality which lend themselves readily to caricature.

The following is Sir Charles Lyell's graphic account of this celebrated lecture before the Association at Oxford. Writing to Mantell, June 1832, he says :—

"Buckland was really powerful last night on the megatherium—a lecture of an hour before a crowded audience: only standing room for a third. Lots of anatomists there ; paper by Clift; the gigantic bones exhibited, and

AWFUL CHANGES!

MAN FOUND ONLY IN A FOSSIL STATE; REAPPEARANCE OF
ICHTHYOSAURI.

"A change came o'er the spirit of my dream."—BYRON.

A LECTURE.—"You will at once perceive," continued Professor Ichthyosaurus, "that the skull before us belonged to some of the lower order of animals; the teeth are very insignificant, the power of the jaws trifling, and altogether it seems wonderful how the creature could have procured food."

still to be seen there, but likely to be removed by-and-by. Buckland made out that the beast lived on the ground by scratching for *yams* and *potatoes*, and was covered like an armadillo by a great coat of mail, to keep the dirt from getting into his skin, as he threw it up. As he was as big as an elephant, the notion of some that he burrowed underground must be abandoned. 'We may absolve him from the imputation of being a borough-monger; indeed, from what I before said, you will have concluded that he was rather a radical.' He concluded with pointing out that as the structure of the sloth was beautifully fitted for the purpose for which he was intended, so was the megatherium for his habits. 'Buffon therefore, and Cuvier even, in describing the sloth, and Cuvier the megatherium, as awkward, erred. They are as admirably formed as the gazelle,' etc. It was the best thing I ever heard Buckland do." [1]

In the possession of the writer is the original manuscript from which Buckland gave the lecture, written out for him in his wife's clear handwriting. From this document a few extracts may be given which will show the careful manner in which he arrived at the habits, form, and character of this monstrosity, whose fossil remains arrived at such a very opportune time.

"It has occurred that within these few days there has arrived in London a large portion of an animal apparently the most monstrous of the monster kind, an animal of which one fragment only had, till within the last few days, ever reached this country. The fragment to which I allude has been for several years in the Ashmolean Museum, to which it was presented by his late Royal Highness the Duke of York—this a portion, and an unimportant portion, of the skeleton of the animal whose entire restoration you there see, [2] a restoration not founded

---

[1] "Life of Sir Charles Lyell," vol. i., p. 388.

[2] A very fine skeleton of the megatherium is to be seen in the Natural History Museum, Cromwell Road. It is a cast, while that at the College of Surgeons is partly Sir Woodbine Parish's original specimen placed there in 1832 and partly a restoration. Dr. Buckland took the greatest pains and interest in setting up the bones, and persuaded Sir Francis Chantrey, one of his oldest and most intimate friends, to allow casts of them to be taken in his foundry. From his friend Dr. Clift's

on imagination, not founded on the putting together of many and various dislocated fragments discovered at distant times and intervals, but founded on one entire animal disinterred from the alluvial districts in the neighbourhood of Buenos Ayres.[1]

"The history of the megatherium—in plain English, 'Great beast'—is very remarkable. It is most nearly allied to the family of the sloth, whose structure is very anomalous, and has been misunderstood by almost every naturalist, including Buffon, and even the immortal Cuvier himself. . . .

" I will illustrate by one example, the specimen before us, the method of investigation, which Cuvier has pointed out and followed, that beautiful and simple method of investigating the structure of every animal, whether of this world or of the past. The system of Cuvier is to begin with the parts that are most important, first with the head

---

anatomical knowledge he says he derived most important aid in his investigation of the animal, and this gentleman's beautiful drawings of the teeth and head were used on the occasion of the lecture. Dr. Buckland begged his audience to judge of the " gigantic " size of the pelvis by the following fact, that Mr. Clift, loaded with all his honours, passed bodily through it, " so that he has come a second time into the world through this cavity in the pelvis of the megatherium ! "

[1] This enormous animal, the megatherium, had been brought to England by Woodbine Parish, his Majesty's Consul at Buenos Ayres. It was discovered by a peasant, who, passing the river Salado in a dry season, threw his lasso at something he saw half covered with water, and dragged on shore the enormous pelvis of the animal; the rest of the bones were obtained by turning aside the current by means of a dam. The animal was about eight feet high and twelve feet long, and its teeth, though ill adapted for the mastication of grass or flesh, are wonderfully contrived for the crushing of roots. The fore feet, nearly a yard in length, were armed with three gigantic claws, each more than a foot long, and forming most powerful instruments for scraping roots out of the ground. The most curious history is that of the megatherium, with his double skull, like a fireman's helmet — See "Bridgewater," 3rd edition, p. 144.

and teeth, then to go downwards through the neck and back to its extremity in the tail. . . .

"In all animals the teeth are indicative of the character of the animal ; and next to the teeth the feet, and in the feet the claws, are indicative of the character : therefore, if we have the teeth and the feet alone, we are able at once to see, in the absence of all other parts, the class and genus of the animal whose teeth and feet we possess.

"We have before us a gigantic quadruped, which at first sight appears not only ill-proportioned as a whole, but whose members also seem incongruous and clumsy, if considered with a view to the functions and corresponding limbs of ordinary quadrupeds. Let us first infer from the total composition and capabilities of the machinery what was the general nature of the work it was destined to perform ; and from the character of the most important parts, namely, the feet and teeth, make ourselves acquainted with the food these organs were adapted to procure and masticate ; and we shall find every other member of the body acting in harmonious subordination to this chief purpose in the animal economy. In some parts of its organisation this animal is nearly allied to the sloth, and like the sloth presents an apparent monstrosity of external form, accompanied by many strange peculiarities of internal structure. . . .

"The megatherium affords an example of most extraordinary deviations and of egregious apparent monstrosity ; a gigantic animal exceeding the largest rhinoceros in bulk, and to which the nearest approximations that occur in the living world are found in the not less anomalous genera of sloth, armadillo, and chlamyphorus ; the former adapted to the peculiar habit of residing upon trees ; the two latter constructed with unusual adaptations to the habit of burrowing in search of their food and shelter in sand ; and all limited in their geographical distribution, nearly to the same region of America that was once the residence of the megatherium.

"The bones of the head most nearly resemble those of a sloth. The anterior part of the muzzle is so strong and substantial, and so perforated with holes for the passage of nerves and vessels, that we may be sure it supported some organ of considerable size ; a long trunk was needless to an animal possessing so long a neck ; the organ was probably a snout, something like that of the tapir, sufficiently elongated to gather up roots from the ground ; such an apparatus would have afforded compensation for the absence of incisor teeth and tusks.   Having no incisors, the megatherium could not have lived on grass ; the structure of the molar teeth shows that it was not carnivorous.

" The composition of a single molar tooth resembles that of one of the many denticules that are united in the compound molar of the elephant ; and affords an admirable exemplification of the method employed by nature, whereby three substances of unequal density, viz. ivory, enamel, and crusta petrosa, or cæmentum, are united in the construction of the teeth of graminivorous animals.   The teeth are about seven inches long, and nearly of a prismatic form. The grinding surfaces exhibit a peculiar and beautiful contrivance for maintaining two cutting wedge-shaped salient edges in good working condition during the whole existence of the tooth ; this is the principle of the mechanism which is adopted in the graminivorous animals.   The various edges are always·kept up ; inside and outside them are depressions, the consequence of which is that the state of the large tooth is in the state of a millstone kept sharp by doing its work ; therefore I say it is the perfection of machinery to keep itself in the highest order by doing the hardest work.   The same principle is applied by toolmakers for the purpose of maintaining a sharp edge in axes, scythes, bill-hooks, etc.   An axe or bill-hook is not made entirely of steel, but of one thin plate of steel inserted between two plates of softer iron, and so enclosed that the steel projects beyond the iron along the entire line of the cutting edge of the instrument : a double advantage results from this contrivance ; first, the instrument is less liable to fracture than if it were entirely made of the more brittle material of steel ; and secondly, the cutting edge is more easily kept sharp by grinding down a portion of exterior soft iron, than if the entire mass were of hard steel. By a similar contrivance, two cutting edges are produced on the crown of the molar teeth of the megatherium. As the surfaces of these teeth must have worn away with much rapidity, a provision, unusual in molar teeth, and similar to that in the incisor teeth of the beaver and other rodentia, supplied the loss that was continually going on at the crown, by the constant addition of new matter at the root, which for this purpose remained hollow and filled with pulp during the whole life of the animal.

" His teeth indicated a peculiarity of structure : they were not calculated to eat leaves or grass ; they were not calculated to eat flesh ; he was an eater of vegetables. What then remained for him but roots ? He has a spade and he has a hoe and a shovel in those three claws in his right hand. You have seen a bull enraged or a dog scratching the ground ; these arms would give the action of a dog or a bull to this animal with a claw such as that, such expanding claws as you now see in this animal of South America in some degree. He is the Prince of sappers and miners. I speak it in the presence of Mr. Brunel, the Prince of diggers. Mr. Brunel eyes him and says, ' I should like to employ him in my tunnel.' ' No,' say I, ' he is not a workman for you ; he is not a tunneller ; he is a canal digger, if you please, so I pray you give him the first job you have to do!' He will not go an inch below a foot and a half : he would dig a famous gutter ; he would drain all Lincolnshire in the ordinary process of digging for his daily food. If you could get him to march in a straight line in the Cambridgeshire fens, he would dig a gutter of incomparable utility. . . . I know from experience the pain of digging in the position in which that animal stood to dig ; the construction of the human form is such that a position on all fours, digging with your own claws, as I have often done, at the bottom of caves, is a very painful thing, and there is a dreadful coming on of lumbago after a quarter or half an hour's work. Now, though the megatherium was digging from morning to night, he never could have tired ; he might go on for ever ; he stood on

three legs as easily or more easily than other animals stand
on four.   The whole structure of his posterior extremities
was rigid ; the pressure was all perpendicular ; he stood
like the poles of a scaffold—there was no muscular exertion
to keep them in their place, therefore he was never
fatigued.

"I say he fed on potatoes ; he lived in those sandy
barren plains of the Pampas where you have roots of that
description.   If his potatoes had been planted by nature
more than one foot and six inches deep in the earth, he
would have starved before he could have got them.   Potatoes, as every one knows, grow from two inches to one foot
below the surface of the earth ; therefore I say the capacity
of his engine for digging and delving shows the depths of
the soil where these roots grew which formed his food.
We find in addition to that nose a snout, a little longer
than that which the tapir now has.   The snout would
pick up the food as the tapir does, and would put it in
as the elephant puts in his apples, and with those sixteen
pegs, as they are contemptuously and sillily called, those
beautiful engines which keep themselves constantly set, he
would munch and munch till he was satisfied.   His business was to be a gardener, a digger, and culler of simples ;
he was a digger up of potatoes and other roots ; he stood
still in dignified composure, and all his concern with other
animals was to keep himself from their annoyance ; he
troubled not them, and woe be to the least beast who
dared to trouble him !"

This account of the first meeting under the presidency
of Dr. Buckland may be fitly closed with the concluding
words of his address :—

"I congratulate each individual here present on the
attainment of what I consider almost the highest beati-
fication of which we are capable in our present state—the
attainment of that personal knowledge and familiar intercourse which this meeting affords with those whose kindred
minds and congenial pursuits have been long familiar to

us through the medium of their works ; a meeting in which they whose heads and hearts we have from a distance long esteemed and loved and venerated are thus brought close together in friendly and brotherly association, and permitted (though but for a short, yet most delightful and intellectual week of their existence) thus to hold sweet counsel and communion together amid these our palaces of peace."

Professor Sedgwick, who required so much persuasion to attend the meeting, seems from the following speech to have enjoyed himself. In thanking Dr. Buckland for the delightful manner in which he had presided over the meeting, he says :—

" All who have witnessed the exercise of his great powers combined with extraordinary tact and temper, so that through his governing influence the jarring elements of a society not yet organised had been brought to order and harmony, must have been struck with admiration. During the long Philosophic banquet of which they had been partaking while in his presence, all seemed to have been living in intellectual sunshine. He looked forward with confidence and pleasure to the results of this union between men of common feeling and common sentiments, who possessed one common object, the promotion of truth and the improvement of mankind."

The British Association, after meeting in the four University cities of the United Kingdom, selected Bristol for the place of its next assembly, in 1836. Buckland was President of the Geological Section. The paper read on Wednesday, August 24th, was on " Saurian Remains," and in the course of the discussion which followed, Buckland mentioned the valuable collection at the Hotwells, which had been of great service to him in the preparation of his Bridgewater Treatise. He also produced a human rib as a geological puzzle. It was filled with lead, and his explanation of the problem, which was abandoned by the learned company as insoluble, was that it belonged to one of the "unfortunate beings who perished at the Custom House at the riots in this city. The animal matter has been roasted out of it by intense heat, and the cavities have been filled with lead."[1]

During his stay at Bristol, Buckland was the guest of the father of Miss Caroline Fox. The young lady writes, August 31st, 1836 :—

"We were returning from the British Association Meeting, and Dr. Buckland was an outside *compagnon de voyage*, but often came at stopping places for a little chat. He was much struck by the dearth of trees in Cornwall, and told of a friend of his who had made the off-hand remark that there was not a tree in the parish, when a parishioner remonstrated with him on belying the parish, and truly asserted that there were seven."

This meeting of the Association at Bristol also finds mention in the Life of Mary Carpenter.

"She entered," it is said, "with alacrity into all the preparations made to receive the *savants* at this meeting. The acquaintance then begun with many distinguished men who gathered at her father's table, was occasionally renewed afterwards. 'In the afternoon,' she wrote in October, ' Professor Buckland called on his way back to Oxford. He stayed half an hour, conversing in a most agreeable and sensible manner about his book,[2] and the contested point of the Creation ; he very wisely determines not to attempt to reason with those who shut their eyes and say that the geologists invent facts. With regard to

---

[1] *Bristol Gazette*, September 1st, 1836.
[2] The Bridgewater Treatise.

the progressiveness of the Creation, proved by geologists, he remarked : " Let man be placed in the early periods of the earth ; deprive him of oxen, horses, and all domestic animals " (you know that none are to be found in the limestone) ; " put him to live among the crocodiles and mammoths, and he would die." ' " [1]

In a course of Bampton Lectures preached at St. Mary's, Oxford, in 1833, the British Association was attacked as mischievous and absurd ! The attack induced Buckland to write to Mr. W. Vernon Harcourt :—

" In my humble opinion it is highly expedient for the interests of the Association and of the University that you should take up the subject in a manner which no man can do as well as yourself, to set the question at issue before the public on its right footing."

The attitude which was assumed by many theologians towards science, and especially towards geology, was at this time exceedingly hostile. Nor did the Professor escape attack. " Buckland is persecuted," writes Baron Bunsen to his wife in April 1839, " by bigots for having asserted that among the fossils there may be a pre-Adamite species. 'How!' say they ; 'is that not direct, open infidelity ? Did not death come into the world by Adam's sin ?' I suppose then that the lions known to Adam were originally destined to roar throughout eternity ! "

It was about this time that Buckland was asked by the rector of the parish in which William Smith was born, if the geologist was not an ignorant old humbug. On another occasion, when he was stating some geological

---

[1] "Life and Work of Mary Carpenter," by J. Estlin Carpenter.

truths concerning saurians to a hunting and shooting clergyman near Lyme, in whose parish they abound in the quarries of lias, his sporting friend stopped the Professor by saying, "'Tis very well for you to humbug those fellows at Oxford with such nonsense ; but we know better at Mugbury !" "Such is the honour of prophets in their own country!" adds the Professor.

Among the numerous foreigners whom common tastes, interests, and pursuits made known to Buckland was Professor Agassiz. Their life-long friendship began in 1834, when Agassiz was Buckland's guest at Christ Church, and was received by scientific men with a cordial sympathy which left not a day or an hour of his sojourn in England unoccupied. Dr. Buckland writes to Agassiz, August 1834 :—

" I am rejoiced to hear of your safe arrival in London, and write to say that I am in Oxford, and that I shall be most happy to receive you and give you a bed in my house if you can come here immediately. I expect Monsieur Arago and Mr. Pentland from Paris to-morrow (Wednesday) afternoon. I shall be most happy to show you our Oxford Museum on Thursday or Friday, and to proceed with you towards Edinburgh. Sir Philip Egerton has a fine collection of fossil fishes near Chester, which you should visit on your road. I have partly engaged myself to be with him on Monday, September 1st, but I think it would be desirable for you to go to him on Saturday, that you may have time to take drawings of his fossil fishes.

" I cannot tell certainly what day I shall leave Oxford until I see M. Arago, whom I hope you will meet at my house on your arrival in Oxford. I shall hope to see you Wednesday evening or Thursday morning. Pray come to my house in Christ Church with your baggage the moment you reach Oxford."

Agassiz always looked back with delight on his first visit to Great Britain. Guided by Buckland, to whom not only every public and private collection, but every rare specimen in the United Kingdom, seem to have been known, he wandered from treasure to treasure. Every day brought its revelations, until, under the accumulation of new facts, he almost felt himself forced to begin afresh the work which he had believed to be well advanced. He might have been discouraged by a wealth of resources which seemed to open out countless paths, leading he knew not whither, but for the generosity of the English naturalists, who allowed him to cull, out of sixty or more collections, two thousand specimens of fossil fishes, and to send them to London, where, by the kindness of the Geological Society, he was permitted to deposit them in a room in Somerset House. The mass of materials once sifted and arranged, the work of comparison and identification became relatively easy. He sent at once for his faithful artist, Mr. Dinkel, who began, without delay, to copy all such specimens as threw new light on the history of fossil fishes, a work which detained him in England several years.[1] On October 13th, 1834, Dr. Buckland writes to Mr. Vernon Harcourt : " My fishing excursion with Agassiz has ended most prosperously ; he has caught too large a multitude for his publication without expansion beyond its already bulky dimensions." Again in 1835 he writes : " I hope you will be able to get Agassiz another grant of £100 in consideration of his labours." In a letter to Murchison, written in the same year, he says : " Harcourt seems to agree in

---

[1] " Louis Agassiz : His Life and Correspondence."

the propriety of asking for another grant to Agassiz, if he
brings with him some good work done out of English
Fishes since last meeting." He knew well the hard struggle
Agassiz had with life, and in his generous large-heartedness
did all he could to assist him. Both Dr. Buckland and his
wife had the greatest affection and esteem for the simple-
minded young Swiss Professor, and both cordially sympa-
thised with him in his enthusiastic love of science, as well
as in his belief that scientific facts are in truth but transla-
tions into human language of the thoughts of the Creator.
For such aid and sympathy Agassiz was deeply grateful.
His private letters contain touching passages, in which,
in the most natural manner possible, his enthusiasm breaks
to the surface, or he regrets his want of money, not for
himself, but for the work he longed to complete, or
expresses his heartfelt gratitude towards all those who
helped him to bring out his splendid addition to the science
of geology.

Not only in his indefatigable energy in the cause of
science, but also in his forgetfulness of his own domestic
comfort, Agassiz greatly resembled Buckland. It might
have been Buckland, if it had not been Agassiz, who—
if the story be true—prepared his coffee in the morning
and his tea in the evening in the same saucepan in which
all day he was boiling up specimens for skeletons! The
two friends were alike also as lecturers and founders of
museums and scientific societies, and in their power of
communicating to others their own enthusiasm. Change
Neuchâtel into Oxford, and the following description of the
Swiss Professor might have been written of Buckland :—

" Il se montra, dès ses débuts, un professeur merveilleux,
qui communiquait sa flamme à ses auditeurs et les entraî-
nait à la conquête de la verité. Il se produisait sous
son influence un remarquable mouvement scientifique, que
les événements de 1848 devaient brusquement interrompre,
de cette époque datent la fondation de la Société des
Sciences Naturelles et l'extension du riche musée."

Agassiz was fired with the same love and passion for his
museum which inspired Buckland. The Swiss Professor
had many brilliant offers of advancement, but " ce que le
retenait surtout en Amérique, c'était son musée de Cambridge, sa création, l'œuvre de sa vie en faveur de laquelle
il avait su éveiller l'intérêt général." [1]

Though the two men were equal in their love for their
respective museums, they were not equally fortunate in
obtaining recognition of the value of their collections.
The appreciation of the new geological science by Americans in 1858 forms a striking contrast to the neglect of
it which was evinced by the Oxford University in 1856.
The legislature of Massachusetts gave a valuable site to
Agassiz for his museum ; a private individual bequeathed
50,000 dollars ; and private subscriptions were raised
which amounted to 71,000 dollars. Buckland's museum,
on the other hand, which was the result of forty years
of travel, toil, and self-denial, has almost perished for want
of a few hundred pounds from the University chest to
unpack and arrange it in the new building to which it
was removed in 1856.

Agassiz was at this time engaged in working out his

---

[1] " Louis Agassiz "—Philippa Godet.

glacial theory, and found in Buckland an uncompromising opponent. " We have made," writes Mrs. Buckland in 1838 to the Swiss Professor, " a good tour of the Oberland, and have seen glaciers, etc., but Dr. Buckland is as far as ever from agreeing with you." It is no slight proof of his openness of mind that he frankly acknowledged his error, when he found that the discoveries of Agassiz satisfactorily explained the existence of boulders and large water-worn stones in positions far above what is now the reach of the agencies to which they must have been at one time subjected. To complete the glacial theory, the two friends travelled together to Aberdeen to confer with the celebrated Professor Fleming, to whom in his monograph on Fossil Fishes Agassiz refers. " We have found," writes Buckland in 1840 to the Aberdeen Professor, " abundant traces of glaciers round Ben Nevis." To the glacial theory he became an enthusiastic convert, and was not satisfied till he had made other leading geologists recognise the importance of the discovery. " Lyell," he writes to Agassiz, " has accepted your theory *in toto* ! ! On my showing him a beautiful cluster of moraines within two miles of his father's house, he instantly accepted it as solving a host of difficulties that have all his life embarrassed him." Buckland himself supported Agassiz with an elaborate paper of observations on the polished, striated, and furrowed surfaces of the sides of mountains. In writing to tell his Swiss friend that the paper was being prepared, he adds : " I expect Murchison will be converted by the inspection of the moraines near Lyell's house. I have found similar polish and scratches on the rock of Edinboro' Castle, and

have sent an artist to daguerrotype them." Among the many
diagrams and drawings left by him to Oxford University
is an interesting sketch of the primitive habitation—a block
of schist on the great moraine—in which Agassiz lived on
the Aar glacier. It was known in the scientific world by
the name of "L'Hôtel des Neuchâtelois," and on the sketch
Buckland has written the words "Given me by Agassiz."

Buckland's championship of the glacial theory was the
subject of a poetic "Dialogue between Dr. Buckland and
a Rocky Boulder," written by his friend Philip Duncan.
The following are the lines :—

"Buckland, *loquitur.*

"Say when, and whence, and how, huge Mister Boulder,
And by what wondrous force hast thou been rolled here?
Has some strong torrent driven thee from afar,
Or hast thou ridden on an icy car?
Which, from its native rock once torn like thee,
Has floundered many a mile throughout the sea,
And stranded thee at last upon this earth,
So distant from thy primal place of birth ;
And having done its office with due care,
Was changed to vapour, and was mixed in air.

"Boulder, *respondit.*

"Thou great idolater of stocks and stones,
Of fossil shells and plants and buried bones ;
Thou wise Professor, who wert ever curious
To learn the true, and to reject the spurious,
Know that in ancient days an icy band
Encompassèd around the frozen land,
Until a red-hot comet, wandering near
The strong attraction of this rolling sphere,
Struck on the mountain summit, from whence torn
Was many a vast and massive iceberg borne,

And many a rock, indented with sharp force
And still-seen striæ, shows my ancient course ;
And if you doubt it, go with friend Agassiz
And view the signs in Scotland and Swiss passes."

How effective was Buckland's support of the views of the young Swiss Professor is shown by the testimony of men who were afterwards eminent in science. Thus Professor Prestwich writes : " I was a young man during Dr. Buckland's latter years, and used to listen at the Geological Society to his vigorous and successful advocacy of a glacial period." So, too, Professor Bonney declares that " it is to Dr. Buckland we owe the recognition of the action of glaciers in the country."

Of the glacial theory of Professor Agassiz, Buckland gave the following description in an Ashmolean lecture :—

" Agassiz considers the glacial period was between the ancient and present state of our planet, and that the melting of this ice was the cause of enormous deluges, which have produced in the low lands many of the great accumulations of gravel to which the name of diluvium has been applied. There is abundant evidence of the effects of glaciers (such as now exist in Switzerland) in most of the valleys proceeding from the higher ranges of mountains in this country. It is well known that large pieces of rock constantly fall on the glacier margin. The progressive motion of the ice carries these stones partly on the surface and partly at the bottom of the glacier. Every glacier is thus thickly set with fragments fixed firmly in the ice, like the teeth of a file, and these being slowly forced against the sides and bottom of the valley are continually producing a series of scratches and grooves upon the rocks they pass. The expansive force that moves the glacier is caused by the successive thawings and freezings of the ice. Precisely similar effects appear to have taken place in the north of England and in Scotland in the period before our epoch. That the glaciers in Switzerland once occupied a very different level from their present one, is evident from the fact that the surface of the valley of the Arve, descending from the Grimsel, shows scratches several thousand feet above the level of the present glacier. I found similar scratches over a breadth of two or three miles in a high valley on the north shoulder of Schiehallion in Perthshire ; the scratches are most distinctly preserved on the surface of two dykes of porphyry, but are also apparent on the harder kinds of slate stone. Traces of the action of glaciers down to the present level of the sea are distinctly visible, between high and low water mark, upon the surface of the granite on the left margin of Loch Leven, and also at Bunaw Ferry on Loch Etive.

"Large round blocks of granite are brought down by the glaciers. Similar cases of transport by ice occur at the present time in the Arctic regions, where vast masses of stone and mud are drifted annually to sea on icebergs, to be stranded on distant shores. In this way we can explain the condition of the east coast of England, where blocks of Scandinavian porphyry have been stranded by icebergs from the Baltic."

The parallel roads of Glenroy may also, Buckland says, be satisfactorily referred to a lake produced by two glaciers descending from the north and east sides of Ben Nevis to the valley of Glen Spean. He also discovered similar signs of glacial action in Wales. Mr. Murray Browne says that—

"Dean Buckland pointed out distinct traces of glacier action at Aberglaslyn Pass, near Beddgelert. This was (I believe) the first time that traces of glacier action were pointed out in Wales." (Every one can now see them at every corner.) "The Dean made a note to this effect in the visitors' book at the Goat Hotel, Beddgelert. The page in which this was written was cut out and framed, and for many years hung up in the hotel. Recently, owing to a

Scratched in a Glacier Thirty three
thousand three hundred
Thirty Three Years before

Scratched by a Cart
Wheel on Waterloo
Bridge the
day before
yesterday

Prodigious
Glacier
Scratches

Scratched by T. Sopwith

The Rectlinear Course of these
Grooves corresponds with the
motions of an IMMENSE
BODY the momentum of which
does not allow it to change its
Course upon Slight Resistances

COSTUME of   the GLACIERS

[Face p. 145.

change of ownership, it has passed into other hands, and is now, I believe, in the possession of a Mr. Jones of Mallington, Chester, who owns property in Beddgelert. The Dean was ridiculed about it at the time, as mentioned in Frank Buckland's Life."

Mr. Sopwith, who was Dr. Buckland's companion in some of his tours in search of glacier scratches, made a semi-caricature of Buckland, who, encumbered with the numerous heavy cloaks, thick travelling boots, bags of fossils, and rolls of maps, presents a figure fancifully like a glacier. The sketch is entitled "The costume of the glacier." Dr. Buckland is represented as standing on a smooth bit of rock covered with scratches under his feet, and the explanation is then given : " The rectilinear course of these grooves corresponds with the motions of an immense body, the momentum of which does not allow it to change its course upon slight resistance." By his side are drawn " specimen No. 1, scratched by a glacier thirty-three thousand three hundred and thirty-three years before the creation ; No. 2, scratched by a cart wheel on Waterloo Bridge the day before yesterday ; the whole picture being scratched by T. Sopwith."

Of Buckland's English friends, the Archbishop of York (Vernon Harcourt) and his family were among the greatest, and, of course, when the Archbishop came to reside at Nuneham, near Oxford, the intimacy between the families greatly increased. From the Harcourt papers, lately published for private circulation only, it is interesting to read the correspondence which the Archbishop carried on with most of the illustrious people of the day, both scientific and political. It was the Rev. William Vernon Harcourt at

whose instigation Buckland discovered in the cave at Kirkdale in 1821 the fossil remains which became the nucleus of
the York Museum,[1] and it was also he who, ten years afterwards, was the mainspring and active secretary of the
British Association in 1831. During the summer term at
Christ Church scarcely a week passed without a party from
Nuneham coming over either to breakfast (breakfast parties
were then the fashion) or to luncheon. The frequency of
such arrivals vividly impressed Dr. Buckland's butler. He
had, he says, "good cause to remember those parties, for
two or three carriage loads would come over at a time,
and eighteen or twenty would sit down to luncheon, and
Master Frank was always sent round the table to show
the guests the Siberian mammoth, which had been
mounted in a silver box. The children always came
down to the drawing-room afterwards, to see the company
before they went off to the Museum."

In June 1841 the Archbishop had the honour of enter-
taining the Queen and the Prince Consort at Nuneham,
during their visit to Oxford. Mrs. Buckland notes in her
journal her regret that, owing to the serious illness of
one of her younger children, she is unable to leave the
house to take part in the enthusiastically loyal reception
given by the citizens and students of the University city
to the Royal visitors. The unusual bustle in the beautiful
streets may be imagined. The Queen and Prince arrived
on the 12th of June, and coaches, flys, tandems, and every

---

[1] This Museum was the origin of the establishment of the Yorkshire
Philosophical Society, of which William Vernon Harcourt was chosen
President.

available vehicle, were filled with anxious subjects, whose
loyalty and curiosity kept pace with each other.   Nuneham
Park, with its velvet lawns sloping down to the Thames,
was thrown open to all by the courtly hospitality of its
venerable owner. The result was that thousands of spectators (and among them the village schoolchildren in their
white dresses) witnessed the arrival of the Royal Party,
escorted by the Oxford Yeomanry. From all sides the
progress of the distinguished visitors was saluted with
tremendous cheering.

On the following day, June 13th, 1841, Prince Albert
drove into Oxford to be present at the annual commemoration in the Sheldonian Theatre.   It was hoped
that the Queen and Prince would both attend it; but
Her Majesty was "dissuaded," says the *Times*, by cogent
reasons from accompanying her Royal Consort. Amongst
these reasons it is sufficient to particularise this one. The
University authorities would have been compelled, by
ancient prescription, to grant an additional vacation of an
entire term, a concession which would have been attended
by great inconvenience.   The Prince Consort was met
at the Schools by the Duke of Wellington, who was
the Chancellor of the University; the Vice-Chancellor,
Dr. Wynter, President of St. John's College; and by the
Heads of Houses, all wearing Court dress. At 10.30 a.m.
the procession, headed by Prince Albert, entered the
theatre. After the proceedings in the theatre [1] were over,

---

[1] Professor Keble delivered the Creweian Oration in Latin. The
prize essays were then recited. The Latin essay was by Benjamin
Jowett, late Master of Balliol.

Prince Albert was taken to the Town Hall, where in the council chamber addresses were presented to him by the City of Oxford and by the County. Immediately after the address the Prince proceeded to St. John's College, a sumptuous entertainment being served up in the hall. The hall doors were thrown open even during the luncheon, and strangers were permitted to view the whole proceedings. Every part of the College exhibited the most boundless hospitality. In fact open house was kept, and vast numbers availed themselves of the opportunity of indulging in College fare.

The present President of St. John's, the Rev. Dr. Bellamy, relates a curious incident connected with the Prince's entry into St. John's quadrangle. In order to provide a better approach, the College authorities had caused an opening to be made from the street through the middle of the wall which bounds the bank opposite the central gate. The bank being higher than the footway in front of the College quadrangle, they caused planks to be placed between it and the College gateway, so as to make an inclined plane. Then they covered the plane, as indeed the whole space between the opening made in the wall of the bank and the President's lodgings, with red cloth, intending that the Queen (whom they hoped to see) and Prince Albert should alight from their carriages at that opening and walk on the red cloth from it to the President's door. When, however, the arrival took place, the Duke of Wellington's carriage drove up first, and to their horror, for they feared the planks would give way, went straight through the opening and down the inclined plane over the red cloth into the College. The Prince's carriage and the other carriages followed ; but no ill consequences ensued, the planks, though only intended for pedestrians, supporting bravely the weight of the carriages.

After the luncheon the Prince and his party visited the chief objects of attraction in the University, including Buckland's Museum, and afterwards attended Divine Service in the Chapel of New College.

CHAPTER VI.

DR. BUCKLAND'S NOTES ON DRAINAGE; THE AGRI-
CULTURAL SOCIETY AT OXFORD IN 1839, AND AT
CAMBRIDGE IN 1840; EXPERIMENTAL FARM AT
MARSH GIBBON; ALLOTMENTS AT ISLIP; LECTURE
ON THE POTATO DISEASE, 1845; DISCOVERY OF
COPROLITES; PARTIES AT DRAYTON MANOR; LORD
PLAYFAIR'S RECOLLECTIONS OF BUCKLAND.

1839—1845.

"Agriculture feeds us; to a great degree it clothes us; without it
we could not have manufactures and we should not have commerce.
These all stand together; but they stand together like pillars in a
cluster, the largest in the centre—and that largest is Agriculture."[1]

IT has been most truly said that Dr. Buckland devoted
all his varied scientific knowledge and experience to
the benefit of his fellow-men. "This craving," to quote
the words of Professor Williamson,[2] "to be useful in

---

[1] Words spoken by Daniel Webster, American orator and statesman,
responding to the toast of distinguished strangers, at the meeting of the
Agricultural Society in Queen's College Quadrangle, 1839.

[2] Professor W. C. Williamson, LL.D., F.R.S., Professor of Botany
in the Victoria University, Owens Coll., Manchester. The biographer
is greatly indebted to this distinguished gentleman for the assistance
he rendered her when she commenced this memoir. Buckland's
shrewdness in the discovery of great talent is to be seen in the

promoting the welfare of the world around him characterised his entire life."

So far back as 1818 the subject of agriculture had
occupied his attention. And, apart from the natural
affinity between agriculture and geology, there were
special circumstances in the condition of the time which
appealed strongly to so ardent a philanthropist. The fall
of prices after the Peace of 1815 produced wide-spread
ruin, and, for the moment, the distress was aggravated by
the displacement of labour which was temporarily effected
by the introduction of machinery into agricultural operations. Riots of agrarian origin were not infrequent;
machine-breaking, rick-burning, and the destruction of the
shops of butchers and bakers, testified to the almost universal distress and discontent in country districts. Buckland's letters illustrate the disturbances, which became,
from 1815 to 1845, a common feature in English rural life.
Thus, in November 1830, he writes to Murchison a letter
from which the following passage is extracted :—

"If it be a very hard-run thing, I shall feel it my duty
to come up to town, and vote for Herschel as President
of the Royal Society; but I shall be very sorry to leave
home on Monday next without a most urgent necessity,
for my wife's father and mother, six miles from here, are in

following letter written by him in 1834 to the learned Professor, then
quite a youth: "I was much gratified at seeing that the Editor
of the *Literary Gazette* took the same view which I have done of your
interesting account of the British Tombs. I am happy to have been
instrumental in bringing before the public a name to which I look
forward as likely to figure in the annals of British Science. I trust
you will not fail to receive in your native town that encouragement
which strangers, as far as their means extend, are ready to proffer you."

hourly expectation of a mob from Abingdon to set fire to their premises, and there are threats of a mob coming into Oxford from the neighbourhood of Benson, and our streets, every night, are on the point of a row between the town and gown.

" My brother-in-law has just come in with seven prisoners, and has lodged them in Oxford Castle for to-night. To-morrow he will take them to the jail at Abingdon, where there was a rescue this morning of seven out of eight prisoners brought in from Hungerford, and a rescue will be attempted to-morrow when the men are taken over from Oxford Castle. Not one soldier is to be found in the land ; and my brother-in-law is fighting with a party of fox-hunters, turned into special constables, and galloping sixty or seventy miles a day during all the past week."

On so kind-hearted a man as Buckland, the agricultural distress, with which his frequent journeys through the country made him unusually familiar, produced a profound impression. Convinced as he was that the true remedy was to be found in improved methods of cultivation, and in the utilisation of all the assistance that science could supply, he endeavoured, both by example and by precept, to help forward the work of agricultural progress.

Among his MSS. lectures and notes[1] are some very interesting remarks on the possibility of reclaiming bogs. It was, perhaps, his oft-repeated maxim, that "there is no waste in nature," which induced him to take endless trouble—to visit, examine, and make notes on all the different bogs, morasses, fens, and marshes in the United

---

[1] The biographer is much indebted to Professor Green for his great courtesy and kindness in allowing her access to these MSS. and Portfolios, which have greatly assisted her in compiling this memoir.

Kingdom, with a view to the possibility of their being rendered useful for agricultural purposes. The question was becoming urgent. The demand for food grew daily greater, and the falling prices and general agricultural ruin which followed the Peace of 1815 reduced the supply. The population of the country was increasing at an enormous rate, while at the same time there was a constant recurrence of bad seasons.

The mass of information which Buckland collected extends over a period from 1818 to 1847, and apparently has never been published. Public attention was by him directed to the commercial good that would attend the draining of marsh lands so as to render them capable of " yielding their increase," and to the improvement which might be effected in the health of the dwellers in fen districts. Among his memoranda occurs such an entry as this : " Within the last few years Seaton fever and Cambridge fever have happily become extinct." On the possibility of reclaiming peat bogs he writes :—

" A mere peat bog, whether wet or drained, is a mass of inert vegetable matter, which, till some method be discovered of exciting putrefaction, must remain unproductive for ever. The plan of burying the surface under a covering of new matter is one which can only be practised in places where the neighbourhood supplies the necessary materials —by sinking a shaft and raising the under stratum, whether clay marl or limestone gravel. This system may be adopted in most of the bogs of the great central belt of Ireland, they being generally based on the limestone rock, and abounding in hillocks and ridges of limestone gravel, for the burning of which the peat supplies fuel. But, in the majority of the mountain bogs, the distance of lime and the exposure of their situation render the capability and advantages of reclaiming them problematical, although their inclined position would much facilitate their drainage. If a bog be covered with a new and artificial crust, without doubt that crust will be productive, or, if the whole substance of the mass be floated off and carried away, the surface that had been buried will naturally be recovered; but the command of water and the necessary materials are so essential to the execution of these expedients that to apply them on the large scale is almost impossible. It is on the edges of large bogs and near the limestone ridges and hummocks that daily encroachments are made upon the bogs of Ireland; but the pieces enclosed are so small, and distributed among such numbers of peasants, whose time is of little value, that the expense, however great, becomes imperceptible. The land that had been covered by the great moss of Kincardine, near Stirling, was regained by the removal of the entire substance of the moss by Lord Kaimes. The same thing has been done on a less scale in a moss near Londonderry. Bogs are not unhealthy. Ireland abounds with lakes and bogs which might be supposed to have some influence on the climate and animal economy of the inhabitants; but it does not appear that it is anywhere unhealthy. No people are more healthy than those who live in the midst of the most extensive and wettest mosses; their atmosphere is entirely free from those putrid vapours which are the constant attendants of more fertile fens and marshes; even the smoke of peat that constantly clouds their cabins is said to be beneficial to the health of the inhabitants. The symmetry and athletic frame of the Irish, their ardent passions and constant flow of animal spirits, which render them always cheerful, often turbulent and boisterous, are to be attributed to that uninterrupted health and vigorous constitution which are derived from the salubrity of the climate.

" Those bogs with which Ireland is in some places overgrown are not injurious to health. The watery exhalations from them are neither so abundant nor so noxious as those from marshes which become prejudicial from the various animal and vegetable substances that are left to putrefy as soon as the waters are exhaled ,by the sun. Bogs are not, as one might suppose, masses of putrefaction ; but, on the contrary, they are of such a texture as to resist putrefaction above any other substance we know of. I have seen a shoe neatly stitched taken out of a bog entirely fresh ; from its fashion it must have been there some centuries. I have seen butter, called rouskin, which had been hid in hollow trunks of trees so long that it was become hard and almost friable, yet not devoid of unctuosity ; the length of time it had been buried must have been very great, as ten feet of bog had grown over it.

"Captain Cook found peat water did not become putrid after being long kept in warm climates. The antiseptic quality of peat is imparted to water in which it has been infused, and extends to all substances that may chance to be buried in it. In the Phil. Trans. for 1747 is an account of the body of a woman found under a moss in Lincolnshire, which from the antique sandals found on her feet had remained there for centuries ; yet the body had suffered nothing by corruption, the hair and nails were fresh as when living, the skin soft and strong, but had acquired a tawny colour—I should rather say, tanned. A human body was found twelve feet deep in the estate of Lord Moira. It was clothed in garments made of hair, and yet, though they must have been buried before the introduction of the use of wool, the body and clothes were no way impaired. A piece of cloth found ten feet deep in a moss at Glassford, Lanarkshire, was perfectly fresh and well preserved. In 1786 a woollen coat of coarse net-work was found in a bog at the depth of seventeen feet. . . .

"Ireland is inferior in fertility to England, because that which is the most productive of all our strata (the red marl) is in the far greater part of Ireland entirely wanting, and because it possesses not such districts as the marsh lands of Cambridge and Lincoln and the south of Yorkshire. It is true that, in passing from London to Holyhead, you see but little of our most prolific strata ; but cross the island from Exeter to Carlisle, and you will scarce pass over a mile of uncultivated ground. Marsh lands are, for the most part, exceedingly unhealthy, from the putrid vapours that are exhaled from their animal and vegetable contents ; the fens of Lincolnshire and Essex and Romney Marsh afforded, before their drainage, too convincing a proof. Where stagnant water that has not been impregnated with moss prevails, nothing is more common than intermittent fevers and malignant epidemics. The soil of the English marshes is a black spongy moor of rotten vegetable matter. The bogs of Ireland consist of inert vegetable matter, covered more or less with unproductive vegetation and containing a large quantity of stagnant water. The difference between these soils is, that the rotten vegetable matter of the one produces unrivalled crops of corn and grass, whilst the inert vegetable matter of the other throws out no kind of plant useful to man.

" In Ireland wood is scarcely ever used for fuel, and the great supply for the poor is from the extensive bogs, which, near great towns, become on this account a valuable property. A bog near Limerick sold for £80 an acre. Much turf is consumed in Dublin and Limerick. . . . The tolls of the Dublin turf boats alone produce an income of £10,000 per annum. The season for cutting, drying, and carrying home turf is considered in Ireland, and many parts of Scotland, to be of as much importance as the harvesting of corn. An idle alarm has often been excited by the plans of draining and reclaiming bogs, as if the stock of fuel would be thus destroyed. But, after a bog has been drained and covered with a thin stratum of earth, the mass below, thus becoming the subsoil, will be so compressed as to afford a better fuel than before, and in a state that requires much less drying. The stock indeed would in this case be exhaustible, whereas at present it replenishes itself. But there can be no question but that it would be a national advantage to convert to the purposes of supplying food to man those bleak, barren, and dreary wastes which now answer no other purpose than that of supplying fuel."

In 1839 Buckland, whose natural talent for organising societies was further excited by his strong conviction of the intimate connection between Agriculture and Geology, succeeded, together with Sir Thomas Acland, Mr. Philip Pusey, and others, in offering a reception at Oxford to the new science of Agriculture as enthusiastic as that which the University City had accorded seven years previously to the sister science of Geology. The founder of the Agricultural Society, and its first President, Lord Spencer, was a keen agriculturist. Dr. Gilbert, the Vice-Chancellor (afterwards Bishop of Chichester), was a great friend of Dr. Buckland's, so there was no difficulty in inducing the University authorities to allow the quadrangle of Queen's College to be roofed over for the reception of the expected visitors. This was considered a wonderful achievement, and it took a fortnight to accomplish. On the evening before the meeting two thousand guests sat down to dinner in the covered quadrangle. Buckland made an eloquent speech, detailing the many advantages which the promoters of the Society hoped would arise from associating practice with science. Many allusions record the effect produced by his address ; but at that time there were no reporters, and, therefore, no connected record is preserved of this and many other speeches. It is curious to see, as has been already noticed, how meagre the reports were, for many years to come, even of large meetings like the British Association. The first day of the meeting finished with another large dinner in the quadrangle, where Sir Thomas Acland proposed the toast of Dr. Buckland, President of the Geological Society, who had done so much for science

in Oxford and in the world at large. In the course of
his reply, Buckland suggested that a joint Committee
should be formed from members of the Agricultural Society
and of the Geological Society, in order to co-operate for
the improvement of agriculture.

"We all," he said, "have to deal with one common
parent, the Earth. It is our business, as geologists, to
consider the history of its origin and the cause of its
present condition ; and it is your business to operate on
the surface, and extract from it the abundant riches with
which Providence has stored it. From such a combination
we may anticipate the most splendid results."

In the following year there was a meeting of the
Agricultural Society at Cambridge, after the precedent
set them by the British Association. In the hall of Trinity
College Buckland again pointed out the great advantages
to be derived by agriculture from the study of geology.
In conjunction with Mr. Murchison and Mr. de la Bêche,
he also undertook a gratuitous survey for the Society ; and
when, in March 1840, the Agricultural Society obtained a
Royal Charter, Buckland was the first honorary member.

In an address at one of the Society's meetings (he
hardly ever missed one) he said :—

"The scientific research for water and the scientific con-
version of barren soils to fertility by the practical application
of geology must obviously be impotent in some of their
most fundamental points without a knowledge of the composition of soil and structure of the earth."

In 1840 Dr. Buckland bought some clay land at Marsh
Gibbon, a few miles from Oxford, in order to try practical
experiments upon draining heavy clay soil. He used to

drive his younger children over with him in a capacious yellow carriage, drawn by a tall, gaunt, gentle horse, called " Old Owen," two or three times a week, in order to superintend the work himself, for it was one of his favourite sayings, " If you want a thing done well, do it yourself." In these early days of drainage and of sanitary building, it was very necessary that the workmen should be instructed at every step. The farmhouse had the foundations of the walls laid in brick, with large slates laid on the top crosswise, as a damp-proof course. Perforated air bricks, made under his direction at the adjacent brickyard, were inserted in the walls, as well as chimney ventilators ; even the stables and cowsheds were also ventilated, though this was considered at the time a very unnecessary waste of money.

In proof of the success resulting from scientific draining and cultivation Dr. Buckland exhibited at the Ashmolean in 1844 an enormous turnip, measuring a yard in circumference, which had been grown on land that before had lain waste.

Marsh Gibbon has retained its fascination for scientific experiment, and has been made by the Ewelme trustees into a model sanitary village, with excellent labourers' cottages let at very reasonable rents. Buckland's work there and his personality are still remembered in the parish, and the Rector, the Rev. Edward Holmes, thus writes to the biographer :—

" As regards the farm which Dr. Buckland sold in 1845, the present owner, Mr. David Jones, says that the drainage pipes were made in two pieces, upper and lower, for main

drainage, and that these were still of benefit on the ploughed
land. Dr. Buckland used soot with the wheat, and rags
used to be ploughed in."

Dr. Buckland was a great favourite among the farmers.
He endeavoured to convey to their minds great facts in
an amusing strain, and was therefore generally successful
in his attempts ; and he was, in a marked degree, a
sympathetic friend and adviser to the labourer, miner,
and mechanic, from whom, as he was wont to say, "he had
learnt many a lesson."

In 1842, as the following extract from a letter written to
Sir H. de la Bêche in November of that year shows, he
was contemplating the purchase of another experimental
farm at Torrington.

"I am going," he writes, "to look at an estate near
Torrington to-morrow with a view to purchasing it, if on
examination it should prove capable of great improvement by thorough draining, sheepfolding, and alternating
crops of green and grain. I fear the climate is bad,
four hundred feet above sea, and within twenty miles
of Dartmoor, over which pass all the south-west winds that
come to Torrington. The whole is barren coal measures.
What think you of their reclaimability by Scotch and
Norfolk husbandry, and of the convertibility of the wet
rushy clay fields into good meadows by thorough draining ?
The climate cannot be worse than Scotland."

The following letter, written by Sir Robert Peel in 1842,
illustrates Buckland's readiness to appreciate and adopt
any agricultural improvement. Smith of Deanston, whose
successful experiments on his Scottish farm revolutionised
the old ideas of drainage, was at this date unknown even

to so energetic an agriculturist as the Premier.   The letter is written to Philip Pusey, himself one of the promoters of the Agricultural Society, and one of the foremost champions of " Practice with Science."

" MY DEAR SIR,—I comply, with the greatest pleasure, with your wish that I should give you the particulars respecting the field which I drained and subsoiled, the produce of which was sent to you by our common friend, Dr. Buckland.

" I was riding with him over a part of my estate in the autumn of 1840.   He remarked a quantity of manure put upon a field, of poor soil, very wet, and in bad condition generally, and said the tenant who placed it there went to very needless expense, for that manure would be of no service while the land remained undrained and in the state in which it then was.   He said also that the land in Scotland, which had been so much improved by Mr. Smith of Deanston, was naturally no better than that on which we were riding, and that in its original state it resembled that land in respect to the quality and properties of the soil in many particulars.

" These remarks of Dr. Buckland did not pass unheeded. I selected the worst field I could find, and determined strictly to follow the plan of Mr. Smith in respect to it, so far as draining and subsoiling are concerned.   I first proposed to the tenant that he should retain the field and do the work under my directions ; but he thought it too expensive for his means, and preferred giving up the field and letting me take it into my own hands.

" Enclosed are the details with respect to the mode of treatment conveyed in answers to queries put by me. The produce you have, I believe, from Dr. Buckland. The weight given is of the turnips, with the tops, but without the fibrous roots.   I was advised by very good practical farmers not to sow turnips, but to have a fallow for wheat ; they thought the land not very well suited for turnips, and that the best period for sowing them was

11

gone by. But I was desirous to exhibit the result of my
experiment, which I had mainly undertaken for the purpose of encouraging others in my neighbourhood to follow
my example.

> "Believe me, dear Sir,
> "Very faithfully yours,
> "ROBERT PEEL.

"WHITEHALL, *January* 13*th*, 1842."

It is scarcely an exaggeration to say that, in 1845,
"famine was sore in the land." The potato disease had
broken out with great virulence, and it was suddenly found
that this crop was as important as the wheat crop. Sir
Robert Peel conferred anxiously with Buckland at this
important crisis, which both these pre-eminently practical
men had foreseen. When in 1845 the potato disease
assumed alarming proportions, Buckland devoted himself
vigorously to the task of ascertaining the causes and
remedies for this severe blow to agricultural prospects.

Having mastered all the facts he could collect by personal experiment, observation, and inquiry, he read a
lecture on the subject before the Ashmolean Society at
Oxford on November 5th, 1845. He afterwards, at very
considerable expense, printed his remarks and distributed
them throughout the whole of England, sending a copy
to the mayor or civil authority of every town, village, and
hamlet. Not only did he give practical advice on the best
means of combating the existing evil ; he also indicated the
substances which formed the best substitutes for potatoes
among the poorer classes, by whom the failure of this
useful vegetable was most severely felt. His exertions in
this cause were fully appreciated, and conferred much

benefit on many who, but for his intervention, would have had no opportunity of obtaining information while there was still time to turn it to practical use.

As potato disease still exists in a more or less degree, Buckland's practical advice for its cure may still be interesting. He says :—

"It is important that all leaves and stems should be burnt, in order to destroy the spawn of the fungi. For next year's planting, small and sound tubers should be selected, and planted whole ; or, if cut, the select parts should be shaken in a sieve with quicklime ; care should also be taken to keep those selected for seed dry."

In the early part of 1846, as soon as he was established at the Deanery of Westminster, he went down to Islip and prospected there for ground suitable for allotments. He chose a piece of land on the top of the hill, overlooking a moor, and well exposed to the sun. This ground was converted, by permission of the Duke of Marlborough, to whom it belonged, into allotments, one of which Buckland rented himself in order to experiment upon growing different sorts of wheat and barley. Greatly to the delight of the tenant and of his whole family, a splendid crop of red-coloured wheat, grown from Egyptian seed, came up, in spite of the bad season, with well-filled ear and tall erect stem, rustling golden red [1] in the summer sunshine, a magnificent advertisement to the Islip labourers of what the earth would grow with care and trouble. At this time any ray of hope that could be held

---

[1] Just the colour of the African gold tribute in the gem room of the British Museum.

out was greatly needed.   Agricultural prospects were at
a very low ebb, and every sort of advice was looked upon
with the utmost contempt and scorn by the John-Trot
geniuses of farming.   The more ignorant a man is, the
more conceited he is; and, in order to convince both
farmer and labourer that science was any good, it was
very important to be able to point to practical proofs of
its benefit.

Buckland's conviction of the immense value which
Geology might confer on Agriculture was abundantly veri-
fied by his discovery of the fertilising qualities of coprolites.
It is difficult for the farmer of to-day, who is provided by
chemistry with numerous agencies to stimulate and enrich
the soil, to appreciate the value and importance of this
discovery.   At that time the farmers' saying "Nothing
like muck" was certainly true, for muck was the only
manure that was available.   No artificial substitutes were
invented, and guano was still unknown except at almost
prohibitive prices.

In Baron Liebig's "Letters on Chemistry" the following
passage occurs, which foreshadows the important results
that have since followed the use of this unexpected source
of agricultural wealth :—

"To restore the disturbed equilibrium of constitution to
the soil, to fertilise her fields, England requires an
enormous supply of animal excrements ; and it must
therefore excite considerable interest to learn, that she
possesses beneath her soil beds of fossil 'guano,' strata of
animal excrements in a state which will probably allow
for their being employed as manure at a very small
expense.   The coprolites discovered by Dr. Buckland (a

discovery of the highest interest to geology) are these excrements ; and it seems probable that in these strata England possesses the means of supplying the place of recent bones, and therefore the principal conditions of improving agriculture."

Speaking of the same valuable discovery, Sir Roderick Murchison recalls, not without a touch of true pathos, the " fervid anticipation " with which Buckland was

" led to hope that these fossil bodies would prove of real use in agriculture ; and one of the many regrets I have experienced since his bright intellect was clouded, was that my friend had not been able to appreciate the truly valuable results that have followed from this his *own discovery*, which, at the time it was made, was treated as a curious but unimportant subject, and almost scouted as being too mean for investigation.   The hundreds of tons of these phosphatic coprolites and animal substances which are now extracted, to the great profit of the proprietors of Cambridgeshire and the adjacent counties, for the enrichment of their lands, is a warning commentary to those persons of the ' cui bono' school who are ever despising the first germs of scientific discovery."

It was the delight of Sir Robert Peel to gather round him at Drayton Manor the most distinguished men of the day in art and science and literature.   From these parties Buckland was hardly ever absent.   He was, indeed, a frequent visitor at Drayton at other times, and both from Sir Robert and Lady Peel he always received the greatest kindness and goodwill.   These parties at Drayton generally consisted of about five or six persons eminent in their various branches of science and information.   The names of George Stephenson, Smith of Deanston, Dr. Lyon Playfair, Baron Liebig, Mr. Mechie, Sir W. Follett, Mr.

Arkwright, Mr. Philip Pusey, Professor Owen, Sir H. de la Bêche, etc., suggest the abundance of the stream of wit and knowledge that must have passed from mind to mind under the worthy presidency of Sir Robert himself. It was at one of these meetings that the following incident, which is recorded in Smiles's " Life of George Stephenson," took place :—

" On one occasion, an animated discussion took place between Mr. Stephenson and Dr. Buckland on one of the great engineer's favourite theories as to the formation of coal. The result was, that Dr. Buckland, a much greater master of tongue-fence than Stephenson, completely silenced him. Next morning before breakfast, when Stephenson was walking in the grounds, deeply pondering, Sir William Follett came up, and asked him what he was thinking about?—'Why, Sir William, I am thinking over that argument I had with Buckland last night. I know I am right, and if I had only the command of words which he has I'd have beaten him.' 'Let me know all about it,' said Sir William ; 'and I'll see what I can do for you.' The two sat down in an arbour, where the astute lawyer made himself thoroughly acquainted with the points of the case ; entering into it with all the zeal of an advocate about to plead the dearest interests of his client. After he had mastered the subject, Sir William rose up, rubbing his hands with glee, and said, ' Now I'm ready for him.' Sir Robert Peel was made acquainted with the plot, and adroitly introduced the subject of the controversy after dinner. The result was that, in the argument which followed, the man of science was overcome by the man of the law ; and Sir William Follett had, at all points, the mastery over Dr. Buckland. 'What do *you* say, Mr. Stephenson ?' asked Sir Robert, laughing. ' Why,' said he, ' I will only say this, that of all the powers above and under the earth, there seems to me to be no power so great as the gift of the gab.' "

It was, however, not so certain that the victory rested with the lawyer.   Frank Buckland says :—

" Although unwilling to spoil a good story, I cannot resist calling Dr. Lyon Playfair into the witness-box to tell his story also. He was present at this very party, and tells me that, although Sir William Follett, armed with practised rhetoric, made a brilliant charge upon Dr. Buckland's theory, yet that the Professor, relying on the stern, stubborn, undisputed facts of geology, and using the weapons of common sense, stood his ground well, honestly, and unshaken in this intellectual assault of arms." [1]

Another story is told of Dr. Buckland and George Stephenson, when both were staying with Sir Robert Peel at Drayton. The party had just returned from church, and were standing together on the terrace near the hall, when they observed in the distance a railway-train flashing along, throwing behind it a long line of white steam. " Now, Buckland," said Mr. Stephenson, " I have a poser for you : can you tell me what is the power that is driving that train ? "

" Well," said the doctor, " I suppose it is one of your big engines ? "

" But what drives the engine ? "

" Oh, very likely a canny Newcastle driver."

" What do you say to the light of the sun ? "

" How can that be ? " asked the doctor.

" It is nothing else," said the engineer : " it is light bottled up in the earth for tens of thousands of years ; light absorbed by plants and vegetables being necessary

---

[1] Memoir, " Bridgewater," 3rd edition, pp. 64, 65.

for the condensation of carbon during the process of their growth, if it be not carbon in another form. And now, after being buried in the earth for long ages in fields of coal, that latent light is again brought forth and liberated, and made to work, as in that locomotive, for great human purposes."

Like a flash of light, the saying illuminated in an instant an entire field of science.[1]

The following letter, written by Lord Playfair to the biographer, recording some memories of Buckland, may be appropriately inserted in a chapter mainly devoted to the Dean's agricultural investigations, since Lord Playfair's brilliant discoveries in chemistry have themselves proved of such infinite service to the scientific farmer.

"MY DEAR MRS. GORDON,—You ask me for some personal memories of your father, Dean Buckland, who was one of the best and dearest friends of my youth. I forget the circumstances of my introduction to him, but it must have been in 1840. I had before that year met him at scientific assemblies, and was an admirer of his scientific books ; but until 1840 I do not think that I knew him personally. However, we were introduced, most probably by our mutual friend Sir Henry de la Bêche. Our acquaintance ripened into a closer and more intimate friendship than appears possible by the relations of a man of world-wide fame, in mature years, with a young Scotch youth who had just emerged from his scientific studies at College. The kindness of Buckland's heart explains this anomaly. I had published one or two original investigations in Germany, which had attracted some attention among chemists, and I found to my surprise that, both at Berlin and in London, the young chemist of

---

[1] "Cyclopædia of Nature Teachings," Hugh Macmillan, LL.D.

twenty-one was welcomed as a colleague by men whose names still remain revered in the history of science. Of these men your father had most influence upon my career, and was certainly my most intimate friend.

" Dr. Buckland had for some time taken much interest in the relations of geology to agriculture, but had found these to be of a complicated character, for, though the rock beneath the soil influenced the crops in a marked degree, it was less dominant in its influence than the surface soil, which frequently consisted of detritus having little relation to the geological structure beneath. This led your father to look to chemistry as a science which might be brought into more useful practical connection with agriculture. Liebig had shortly before written his masterly work on Agricultural Chemistry, which I introduced to this country by an English translation. This made me the natural exponent of Liebig's views in England, specially as I kept myself in close correspondence with my great master, and became acquainted with all his new researches. In these your father was much interested, and did much to popularise them among agriculturists. Personally I did not then see much of your father, as I resided in Clitheroe and afterwards in Manchester; but I visited him in Oxford on two occasions, and we had lively conversations as to the best methods of inducing farmers to throw the light of science on their important industry.

" In 1842 Baron Liebig offered to pay me a visit in Manchester, when I was living in humble lodgings, ill calculated to receive my illustrious friend. On consulting Dr. Buckland he suggested that I should induce Liebig to make a tour in Great Britain, where he was certain to be received with welcome and with honour. Though I became ' personal conductor ' of this tour, your father joined us in part of it and contributed much to its success. We went together to the meeting of the Royal Agricultural Society, which will explain the fact that, in the published print to which you refer, his portrait appears standing beside Lord Ducie, Baron Liebig, T. C. Morton, and myself. This tour, in which your father took such an active part, did a great deal to stimulate the leading agriculturists of the country to carry out the motto of the Royal Agricultural Society, ' Practice with Science.' Among other houses which he visited were those of Sir Robert Peel at Drayton Manor, Lord Ducie at Portworth, Lord Fitzwilliam at Wentworth, Lord Essex at Cassiobury, Philip Pusey, then the ruling spirit of the Agricultural Society, Mr. Webb and Mr. Miles at Bristol, and Mr. Crosse. The opportunity was taken of our visits to hold meetings in the neighbouring towns, and the genial, amusing speeches of your father contributed much to their success. Baron Liebig always spoke in German, so my function chiefly consisted in rendering his speeches comprehensible to the audience, by repeating them in English after he had finished.

"One notable fact should not be omitted. Your father had shown that the coprolites found in various rocks could not be anything but the fossil dung of extinct animals, as the intestinal marks were still obvious. Dr. Buckland took us to see these coprolites in the strata in which they occur. Liebig, on being convinced of their probable origin, said they must contain abundance of phosphate of lime, the most needed manure for our exhausted soils. By the post of the same day I sent some to my laboratory in Manchester, where it was found that they abounded in phosphate of lime. Later, on his return to Germany, Liebig made complete analysis of the coprolites, and what your father termed ' pseudo-coprolites,' which were also found to contain this important earth. This was the origin of the great industry of Superphosphates, which has done so much for agriculture. During part of our tour Dr. Daubeny was with us, and he suggested that mineral phosphates such as he had seen in Estramadura might be used when coprolites failed, and this source is now largely used in agriculture.

"I hope that I have answered your question as to whether your father did much to promote the application of science to agriculture. In relation to this you ask me another question, whether Dr. Buckland and the great Minister Sir Robert Peel worked together for this purpose. To make this clear to you I must interpolate an anecdote of my own personal history. While I was Honorary Professor of Chemistry at the Royal Institution of Manchester, the chair of chemistry at the University of Edinburgh became vacant, and for this I was an unsuccessful candidate, though I was second in the running. While smarting under this disappointment I received a letter from Faraday saying that the University of Toronto in Canada had entrusted him with the selection of a Professor of Chemistry, and that he nominated me. As there was little opening at that time for any chemist in this country, I accepted the appointment. This was a grief to your father, who did not wish me to leave the country. No doubt he represented his views to Sir Robert Peel, for at this time the latter invited me to pay him a visit at Drayton Manor. As I had never seen this great statesman, I was much astonished at the invitation, which of course I accepted with much pleasure. On going to Drayton Manor I found a large party, including your father. Next morning we found that all the neighbouring landlords and farmers met at Drayton Manor, and they were addressed by Dr. Buckland and by myself, as well as by Sir Robert Peel, on the application of science to agriculture. Reporters were present, and these speeches at the time produced an effect on the public. After staying at Drayton Manor for a few days, Sir Robert Peel told me that he had wished to form his own opinion of me, and that he entirely agreed with Dr. Buckland that I should not take a foreign Professorship, offering his powerful influence to get me employment if I resigned it. It is needless to state that I did, and I am proud to say Sir Robert Peel honoured me with his friendship till his death. On my future visits, which were numerous, to Drayton Manor, I generally met your father. On one of these occasions the Deanery of Westminster became vacant, and your father thought that Sir Robert Peel would offer it to him. Though it was not a bishopric, Dr. Buckland had the genuine feeling of 'nolo episcopari,' and thought it was his duty to refuse it.

"For at least an hour in my bedroom I had to combat his scruples of conscience, or rather of his modesty, until at last I got his promise to accept the appointment if it were offered. Other persons more competent than I am will tell you about his life as Dean of Westminster. Not that I am ignorant of it, for there was scarcely a week that I did not dine at the Deanery, and continued in the enjoyment of his friendship till the cloud came over his mind.

"But I may conclude with a short estimate of his character. Dean Buckland was one of the most active-minded men I ever met. To all subjects under his attention he gave the best efforts of his mind. Of course geology was his special science, but he did not limit himself to it. Whenever he thought he could be useful to humanity, he threw himself into the work with heart and soul. He often co-operated with me, for instance, in promoting public health, while I acted as a commissioner to investigate into the sanitary condition of the United Kingdom. He was deeply impressed with the opinion that 'cleanliness is next to godliness,' and he was a most robust preacher on this subject. During the cholera he rather startled the congregation on the Day of Humiliation by preaching on the text 'Wash and be clean,' and an admirable sermon it was. His geniality and love of humour, and even of downright fun, made him a charming companion.

"I need not tell his daughter of the deeper qualities of the man, of his love of truth, of the real reverence of his nature notwithstanding the exuberance of his spirits. His kindly nature few could know better than myself, though I am sure there are many men of science who could testify, as I can, that they owe much to his warm sympathies and active friendship when they were fortunate enough to win it.

"I am, dear Mrs. Gordon,
"Yours sincerely,
"PLAYFAIR."

CHAPTER VII.

LANDSLIP AT AXMOUTH, 1839; BURMESE, AMERICAN, AND
INDIAN COLLECTIONS OF FOSSILS; THE MOA; THE
FOUNDATION OF THE SCHOOL OF MINES; MEETINGS
OF THE SOCIETY OF CIVIL ENGINEERS; INTEREST IN
PISCICULTURE.

ON Christmas Day, 1839, occurred the remarkable landslip at Axmouth, the extent of which, says Buckland,
"far exceeds the earthquakes of Calabria, and almost the
vast volcanic fissures of the Val del Bove on the flanks of
Etna." Dr. and Mrs. Buckland were both quickly on the
spot, and while the Professor made careful investigations
into the cause of the catastrophe, his wife, with her clever
pencil, made a series of careful drawings of this curious
phenomenon, from one of which the illustration on the
following page is taken. Buckland at an Ashmolean
meeting thus describes the event :—

"The recent sinking of the land and elevation of the
bottom of the sea at Axmouth, Devon, which occurred
during two days, December 25th and 26th, have no analogy
to the motions of an earthquake, but come from an entirely
different cause. The cliffs on that part of the coast consist
of strata of chalk and cherty sandstone, resting on a thick
bed of loose sand or fox-mould, beneath which is a series
of beds of fine clay impervious to water. Owing to the

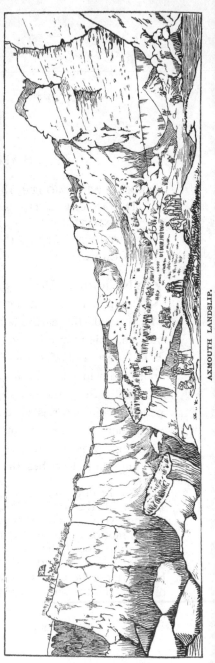

AXMOUTH LANDSLIP.

long continuance of wet weather in the last autumn, the lower region of the fox-mould had become so highly saturated with water as to be reduced to semi-fluid quicksand. The coast from Axmouth to Lyme Regis presents vertical cliffs of chalk about five hundred feet above sea level, between which cliffs and the beach a space, varying from a quarter to half a mile in extent, is occupied by ruinous fallen masses of chalk and sandstone, forming an undercliff similar to that in the south coast of the Isle of Wight. The landslip at Axmouth began in the night of December 24th, 1839, and during the following day slight movements of the undercliff were noticed ; a few cracks also appeared in the fields above.

" About midnight of December 25th the inhabitants of two cottages in the undercliff were awakened by loud sounds produced by the grinding of slowly moving masses of the adjacent rocks ; they found the floors of their houses rising upwards towards the ceiling, and with difficulty escaped. In a few hours one cottage was thrown down. About midnight also the two coastguards observed a huge reef of rocks gradually rising out of the sea at a short distance from the shore ; they moved slowly upward during December 26th, until a reef or breakwater was formed half a mile long and ranging from ten to forty feet in height, between which and the shore was a basin of salt water about five acres in extent and in some parts twenty-five feet deep. The men who saw the reef rising fled to the top of the cliffs, where they soon found the fields on which they trod intersected by chasms, from which they made their escape with difficulty. Fifty acres were gradually severed from the mainland during December 26th. Of these a portion subsided about fifty feet below its former level, and the rest sank into a tremendous chasm extending three quarters of a mile from east to west and varying in breadth from two hundred to four hundred feet. Towards the face of the new cliff, a portion of the mass presents a most picturesque appearance of ruin and confusion, arising from the fact of its having broken up into fragments, which having sunk to unequal depths and being divided by deep chasms give the appearance of castles, towers, and pinnacles. The upward movement of the reef was simultaneous with the downward movement of the land. A similar elevation of a reef was produced in March 1790 by the subsidence of about eight acres of chalk in the parish of Beer, three miles west of Axmouth. A third example of the same kind but on a minor scale took place last February in the day-time at Whitlands, about a mile and a half west of Lyme. The most decisive confirmation of the theory of hydrostatic pressure causing the elevation of reefs beneath the sea was afforded at Whitlands by the rising of two reefs at a short distance from the shore, which were seen to rise as the undercliff descended."

Buckland concluded his address by giving a list of land-slips which have occurred along the coast at various times, and by stating that similar landslips, under similar conditions, often occur on the sides of inland valleys. A stratum of solid stone, resting on a bed of permeable sand, beneath which is a bed of impermeable clay, are the conditions of most of the landslips from the sides of hills into the adjacent valleys.

In the cause of his beloved science a journey to the extreme North of Great Britain was nothing to Buckland. A large artificial lake under the Pentland Hills, some sixty feet deep, had dried up after a season of great drought. As many parts of the bottom of this lake were calculated to throw much light upon several important phenomena in geology, an opportunity occurred of acquir-ing evidence of which Buckland was not slow to avail himself. One of the phenomena which were thus illustrated was the manner in which several species of locomotive fresh-water shells were found congregated in one dense

bed, extending over a small area near the lowest bottom
of the pond to which the water had subsided. Other
beds of sediment at the bottom of the pond were found
crowded with bivalve shells deeply embedded in mud, to
the exclusion of shells of the more locomotive univalves.
These facts seemed to throw light on the appearance in the
Petworth and Purbeck marbles of only one species of uni-
valve shells, and the non-existence of these univalves in
other beds of the same marble which contain exclusively
bivalve shells. Another point was the collection of fish in
one spot. The fish which had survived were congregated
with the surviving molluscs in the remaining shallow water,
and, if this dried up entirely, the first bed of mud formed
by the returning water of the next flood would bury them
in one stratum, after the manner of fish that are entombed
in miscellaneous shoals in the strata of Solenhofen and
other places. Another phenomenon in the pond was the
occurrence of recent footsteps of animals and birds on the
surface of the soft beds of mud and sand since the water
had subsided. These illustrate many similar footprints
which have been discovered upon the slabs of stone in the
new red sandstone formations.

From all parts of the world came specimens and collections, on which Buckland was asked to report. Thus in
1827 he was called upon to examine some fossil animal
and vegetable remains collected by Mr. Crawfurd on a
voyage up the Irawadi from Rangoon to Ava of five
hundred miles. The specimens were principally collected
from a tract of country on the east bank of the Irawadi,
near the town of Wetmasut, about half-way between Ava

and Prome. The bones were found in soil which chiefly consisted of barren sandhills mixed with gravel intersected by deep ravines ; beneath these hills are strata containing shells and lignite, through which wells are sunk about two hundred feet to collect petroleum. Buckland, in his report, suggests that it would be an interesting subject " of inquiry, whether any fossil remains of elephant, rhinoceros, hippopotamus, and hyena exist in the diluvium of tropical climates ; and, if they do, whether they agree with the recent species of these genera or with those extinct species whose remains are dispersed so largely over the témpérate and frigid zones of the northern hemisphere." " It deserves remark," he adds, " that the gavial and several other pachydermata found by Mr. Crawfurd[1] do not now inhabit the Burmese country."

In 1835 another large consignment of specimens arrived from Connecticut and other parts of America. The most important of these were the fossil footprints preserved in sandstone of a gigantic Dinosaur, a link between reptiles and birds, whose feet measured sixteen inches in length exclusive of a large claw measuring two inches. The most frequent distance which intervenes between the larger of these footsteps is four feet ; sometimes they are six feet asunder : the latter were probably made by the animal while running. There were also tracks of another gigantic species, having three toes of a more slender character : these tracks are from fifteen to sixteen inches long, exclusive of a remarkable appendage extending backwards from the

[1] Mr. Crawfurd was the first to find these extinct animals in Asia; most of his specimens are at the Geological Museum, Jermyn Street.

heel eight or nine inches, and apparently intended (like a snow-shoe) to sustain the weight of a heavy animal walking on a soft bottom. The impress of this appendage resembles those of wiry feathers or coarse bristles, which seem to have sunk into the mud an inch deep, while the toes had sunk much deeper. Round their impression in the mud was raised a ridge several inches high, like that round the track of an elephant in clay. The length of the step of this creature appears to have been six feet ; the footsteps on the five other kinds of tracks are of smaller size, and the smallest indicates a foot but one inch long and a step from three to five inches. The length of the leg of the African ostrich, it may be added, is about four feet, and that of the foot ten inches. These tracks appear to have been made on the margin of shallow water, that was subject to changes of level, and in which sediments of sand and mud were alternately deposited.

The next collection came from India, and contained the first specimens ever brought thence to this country. In 1836 Dr. Buckland examined a number of fossils from the hill-slopes and ravines that traverse that part of the Siwalik Sub-Himalayan range of hills which lies between the Jumna and the Sutlej rivers. He describes

"a large ruminating animal called the Sivatherium, approaching the elephant in size, discovered in the Sub-Himalayan range of hills. The jaw of this animal is twice as large as that of a buffalo, and larger than that of a rhinoceros. The front of the skull is remarkably wide, and retains the bony cores of two short, thick, and straight horns, similar in position to those of the four-horned antelope of Hindostan. The nasal bones are salient in a degree without example among ruminants, and exceed

in this respect those of the rhinoceros, tapir, and palæo-
therium, the only herbivorous animals that have this sort of
structure. Hence there is no doubt that the sivatherium
was invested with a trunk, and probably this organ had an
intermediate character between the trunk of the tapir and
that of the elephant."

The Sivatherium is one of a group of a remarkable series
of animals, all of which (with the exception of the giraffe)
are extinct. Buckland pointed out the importance of these
newly discovered fossil animals in filling up intervals in the
order of pachydermata, where links were previously wanting
to connect many living genera, between which the distance
is much wider than in any other species of mammalia.

The famous Moa or Dinornis was also brought before
Buckland. On May 29th, 1843, he read some interesting
letters detailing the discovery of the bones of a gigantic
bird, which must have recently inhabited New Zealand,
even if it did not prove to be still an inhabitant of that
colony. The announcement of its supposed existence
was conveyed in a letter from Dr. Buckland's Torquay
friend, Mr. William Williams, dated February 28th, 1842.
The writer says that, hearing from the natives of an extraordinary monster which inhabited a cave on the side of
a hill near the river Weiroa, he was induced to offer a
reward to any person who produced either the bird or one
of its bones. In consequence of this offer a large but
much worn bone was found, and, shortly after, another of
smaller size was discovered in the bed of a stream which
runs into Poverty Bay. The natives were then induced to
go in large numbers to turn up the mud in the bed of
the same river, and soon brought to Mr. Williams a large

collection of bones, which proved to have belonged to a bird of gigantic dimensions. The length of the large bone of the leg is two feet ten inches. The bones were found a little below the surface in the mud of several other rivers, and in that situation only. The bird to which they belonged is stated to have existed at no very distant period and in considerable numbers, as bones of more than thirty individuals had been collected by the natives. Mr. Williams had also heard of a bird having been recently seen near Cloudy Bay in Cook's Strait by an Englishman, accompanied by a native, which was described to be not less than fourteen or sixteen feet in height, and this creature he supposed to be about the size of that to which the bones belonged. Of these bones, one case had already arrived and a second was daily expected.

A letter from Professor Owen, dated January 21st, 1843, detailed the contents of the box which had arrived ; and from these fragments it was clear that they had belonged to the species of birds which the Professor had already described in the Zoological Transactions [1] from a fragment of the femur which he had received some time previous. The bird forms a new genus, on which Professor Owen bestowed the name "Dinornis Novæ Zelandiæ." His diagnosis of the species, size, and character of the bird was a remarkable testimony to his extraordinary sagacity. By the process of severe philosophical induction, and not by mere guesswork, he was enabled to describe the bird with the utmost accuracy from the inspection of the solitary

---

[1] Vol. iii., p. 32, pl. iii.

small fragment of the thigh which was then the only bone
of the creature in Europe.   The description in every
particular was confirmed by the arrival of Mr. Williams'
specimens.

In all projects which could promote the spread of
scientific and technical knowledge, especially of subjects
connected with his favourite studies, Buckland took a
prominent part.   The foundation of the School of Mines
in Jermyn Street was largely due to his efforts.[1]

Mr., now Sir B. W., Richardson, in his book of extracts
from Mr. Sopwith's Journal, mentions that on June 8th,
1837, "Two events of importance took place : one a visit
to the famous Dr. Buckland, father of the late Frank
Buckland, at Oxford; and a second, the projection of a
School of Mines, arising, as it seems, out of that visit."

"In the breakfast-room," Mr. Sopwith says, "Dr. Buckland
introduced me to Mrs. Buckland and to Dr. Davies Gilbert.
Dr. Buckland said that he had been applied to to recommend some one as a proper person to undertake the office
of Mining Commissioner on the part of the Free Miners.
' I told them,' said the Doctor, ' that they must have nothing
short of Newcastle, and I named Mr. Buddle and yourself.'
I sat next to Dr. Gilbert, and had with him and Dr. Buckland
a conversation on the subject of a School of Mines.   Dr.
Gilbert said that great advantages had been derived from
the institution of a Polytechnic School in Cornwall, of
which he has been an active promoter.   Before leaving,

---

[1] At the suggestion of Sir Roderick Murchison, and at the generous
expense of many of the most eminent scientific men of England, a bust
of Buckland was placed on December 2nd, 1860, in the Geological
Walhalla of the Jermyn Street Museum, in company with the busts of
Sir H. de la Bêche, Professor E. Forbes, Greenhough, Playfair, Smith,
Hutton, and Sir James Hall.

he made me write a minute to the effect that Mr. Buddle and I should dine with him at the Geological Club in London on the following Wednesday."

In the autumn of 1838, at the meeting of the British Association at Newcastle, Buckland again conferred with Mr. Sopwith, Sir Charles Lemon, and others upon the best mode of bringing the subject of an application to Government on Mining Records before the Association. "It was," he said, "indispensable for the country to have a scientific education in connection with manufacture and mining." The immediate outcome of his efforts was a grant of money from the Association for the purpose of collecting and preserving information as to the geological structure and mineral riches of the country. The opportunity was one which was not to be lost. Sections of the strata on the numerous railroads in various parts of the United Kingdom, many of which traversed important mineral districts, were exposed in cuttings, and, before they again became covered, would afford much valuable information.

The collection of all this information in one central spot was one of the objects which Buckland had in view in his projected School of Mines. But the more he considered the scheme, the more varied appeared to be its utility. The Jermyn Street Museum and School was to serve as the central map office of the Geological Society ; as a Mining Record office, where all plans of mines abandoned or existing are registered and kept ; as a statistical office, in which might be collected all the documents that bear upon the mineral produce of the country ; and, finally, as a technical college, where students are trained in mining and assaying. The necessity for such an institution became every day more apparent.

It was not long before his persistent efforts were rewarded. He was able to announce to the Geological Society that he and his friends had obtained the cooperation of the Departments of Woods and Forests and of the Ordnance, of the British Museum, of the Institute of Civil Engineers, and of the British Association, "in furthering and advancing the knowledge of the structure of the earth." He had made out a strong case in favour of such a school when he insisted on

"the pecuniary value and statistical utility of geological investigations in directing the researches of industry to those points where they may be profitably applied, and in preventing such large expenditures of capital as, under ignorance of the internal structure of the earth and the peculiar productions of such geological formation, we have in times past seen thrown away in ruinous searches after coal, when the slightest knowledge of geology would have given information that no coal could possibly be found. Never more shall we witness a recurrence of such unpardonable waste of public money as that which is said to have been lavished in sending lime from Plymouth to build the fortress of Gibraltar, on a rock exclusively composed of limestone."

The projected School of Mines was also to serve as a Museum, in which might be exhibited specimens of the various stones, marbles, and granite which were employed both at home and abroad in building. Buckland had already collected similar samples in the Oxford Museum, in order that ocular demonstration might afford to architects and engineers information respecting the relative durability of building materials which could be supplied in no other way. At the Society of Civil Engineers Buckland was a well-known figure, and on some points a recognised authority. He rarely missed the reading of any paper of importance at the meetings of the Society, taking an active part in the discussion. Whenever his personal aid and influence could be useful, they were cheerfully given. His archæological knowledge was sometimes of great service to the Society. Acquainted with every Roman villa then known in the country, he had not only observed the Roman method of building, draining, and warming their houses, but had also examined the cement in which the beautiful tesselated pavements are so firmly fixed, and had caused models to be made of the peculiar fan-tailed tiles which he discovered at Wheatley Villa, near Oxford. It was a definite article of his archæological creed that Roman villas would not have fallen into ruins so completely, had not snails absorbed the mortar to make their own natural coverings. He constantly, it may be added, brought home from Stonesfield and Wheatley Villas some of the large edible snails that live there; but they did not long survive in the Islip garden.

In London Buckland put his knowledge of the relative durability of different stones to valuable account. He was convinced that the lavish use of Bath stone in the metropolis was a gross mistake, and when Dean of Westminster he would allow none to be used in the Abbey. He preferred Normandy stone or Yorkshire stone, both of which were as cheap and more enduring.

His opinion on this subject was recognised as so authoritative that his advice was often asked by practical men in
important undertakings. The old breakwater at Weymouth
having been much injured by the pholas boring into the
limestone, the engineer consulted Dr. Buckland upon the
best stone to be used in making the new one. After many
exhaustive inquiries, he recommended that Portland stone
(of which St. Paul's is built) should be used, as the
pholas will not bore into it on account of the quantity of
silica or flinty matter it contains. In 1841 he published
a paper in the Proceedings of the Geological Society upon
the agency of land-snails in corroding and making deep
excavations in compact limestone. He examined the
peculiar hollows on the under surface of a ledge of carboni-
ferous limestone rock, and, as he found in them a large
number of the shells of *Helix aspersa*, he concluded that
the cavities had been formed by snails, and that probably
many generations had contributed to produce them. He
intended to ascertain whether the cavities were hollowed
out by these snails by means of an acid secreted by them,
or by means of their rasp-like tongues. In a speech
delivered before the Geological Section of the British
Association at Cambridge in 1845, he discussed the question at some length. The following extract from the
Times gives a summary of what he said :—

" Dr. Buckland described the agency of land-snails in
forming holes and trackways in compact limestone. His
attention had first been called to the subject by a discussion
on the perforations sixty feet high at Tenby Castle, which
were by some taken to be evidence of a raised beach, but

which he considered as the workmanship of land-snails.
He considered that by means of the acid with which they
were provided snails could make perforations into the most
solid forms of limestone, but the perforations were unlike
those made by any other animals, or those made by the
salt of the sea and the carbonic acid of the atmosphere.
These perforations were never found where the rain and
frosts could operate, and always had the aperture downwards. From observations made at Richborough last year,
he had concluded that these perforations were not made
to a greater depth than an inch in a thousand years."

Subsequently he seems to have leaned to the opinion
that the perforations were bored by the rasp-like tongues
of the snails. It was with a view to the establishment or
disproof of this theory that his wife and her youngest
daughter Caroline during his illness at Islip made a large
collection of the tongues of both land and fresh-water
snails, which they mounted in Canada balsam, and careful
drawings were made of them.

Another favourite topic of discussion at the Society of
Civil Engineers was the water supply of large towns.
The following extract from the Bridgewater Treatise on
artesian wells shows the practical value Professor Buckland attributed to this " prime necessary of life," as well as
the poetic view with which he regarded it :—

" In some places application has been made to economical
purposes, of the higher temperature of the water rising from
great depths. In Wurtemberg, Von Bruckmann has
applied the warm water of artesian wells to heat a paper
manufactory at Heilbronn, and to prevent the freezing of
common water around his mill wheels. The same practice
is also adopted in Alsace, and at Canstadt, near Stuttgard.
It has even been proposed to apply the heat of ascending

springs to the warming of greenhouses. Artesian wells
have long been used in Italy, in the duchy of Modena;
they have also been successfully applied in Holland, China,
and North America. By means of similar wells, it is
probable that water may be raised to the surface of many
parts of the sandy deserts of Africa and Asia, and it has
been in contemplation to construct a series of wells along
the main road which crosses the Isthmus of Suez.[1] . . .
Among the incidental advantages arising to man from the
introduction of faults and dislocations of the strata into
the system of curious arrangements that pervade the
subterranean economy of the globe, we may further
include the circumstance, that these fractures are the most
frequent channels of issue to *mineral* and *thermal* waters,
whose medicinal virtues alleviate many of the diseases of
the human frame.

"Thus in the whole machinery of springs and rivers, and
the apparatus that is kept in action for their duration,
through the instrumentality of a system of curiously constructed hills and valleys, receiving their supply *occasionally* from the rains of heaven, and treasuring it up in
their everlasting storehouses to be dispensed *perpetually*
by thousands of never-failing fountains, we see a provision
not less striking than it is important. So also in the
adjustment of the relative quantities of sea and land, in
such due proportions as to supply the earth by constant
evaporations, without diminishing the waters of the ocean;
and in the appointment of the atmosphere to be the
vehicle of this wonderful and unceasing circulation; in
thus separating these waters from their native salt (which,
though of the highest utility to preserve the purity of the
sea, renders them unfit for the support of terrestrial animals
or vegetables), and transmitting them in genial showers to
scatter fertility over the earth, and maintain the never-
failing reservoirs of those springs and rivers by which
they are again returned to mix with their parent ocean;

---

[1] The French have since this time successfully sunk a series of
artesian wells in the Sahara.—R. HUNT.

in all these circumstances we find such evidence of nicely balanced adaptation of means to ends, of wise foresight, and benevolent intention, and infinite power, that he must be blind indeed who refuses to recognise in them proofs of the most exalted attributes of the Creator."

Pisciculture was a subject to which Buckland devoted much attention. It was from his father that Frank Buckland must have inherited his taste for fish-hatching. In 1844 Buckland gave an account of his visit to the experimental ponds at Dumlanrig, in company with Professor Agassiz, who was himself conducting a series of analogous experiments on the trout of the lake of Neuchâtel. The Doctor alluded to the great probable advantages of hatching the ova in artificial ponds with a view to the preservation of the young fry. In the experiments of Agassiz and Sir F. Mackenzie it was found necessary to feed the fry with the paunches of sheep. The growth of the salmon after it descends to the sea was stated by an old fisherman at Axmouth to average a pound a month, and the fish of the different rivers appear to return to spawn at different periods. The food of the salmon in the sea is probably the jelly-fish, for the stomach has many blend sacs and seems adapted for rapid digestion. Dr. Buckland referred to Mr. E. Forbes' observations on the shelly molluscæ, the young of which when hatched are locomotive, float about with little wings, and perhaps furnish food for the salmon. He alluded also to the advantage of assisting the salmon by staircases, where the falls of rivers are too high to be cleared by a single leap of the fish.

His remarks upon the locomotion of fishes are interesting, as the subject is now happily illustrated at the Brighton

Aquarium. Thus he describes gurnards as " closing their fins against their sides, and, without moving their tails, walking along the bottom, by means of six rays, three on each pectoral fin, which they placed successively on the ground. Their great heads and bodies seemed to throw hardly any weight on these slender rays or feet, being suspended in water and having their weight further diminished by their swimming bladder." A flagstone, to which he gave the name " Ichthyopatolites," was sent to the Professor from a coal shaft at Mostyn, with an impression on it like the trackway of a fish, crawling along the bottom by means of its fins.

The late Rev. Gilbert Heathcote, Sub-Warden of Winchester, used to tell the following story in connection with the Professor's observations on the locomotion of fishes.[1]

Dr. Buckland was Mr. Heathcote's guest at a New College " gaudy " dinner, when a turbot was brought on the table. The Doctor, being at that time interested in the question of the movement of fishes, said, " I should like that fine fellow's head and shoulders to examine—just what I wanted : do you think I can have it ? "

" Certainly," said his host, and, when the fish was removed to the side-table, he offered to have it put on one side for him.

" Thank you," said the Professor, " but I would rather cut it off myself, as I can tell better what I want." So up he jumped, napkin in hand, and in a few minutes returned in great glee, with the coveted specimen in his hand,

---

[1] Mr. G. W. Heathcote had observed that a pike can creep along the grass.

wrapped up carefully in the napkin, which he promised to return. The parcel was thrust in the big outside pouch-pocket worn in the dress-coats of those days.

After the dinner was over, Mr. Heathcote was called upon to propose the toast of Dr. Ingram (the celebrated author of " Memorials of Oxford "). He was a junior Fellow, and was rather taken aback at being thus unexpectedly asked to make a speech, and when he got up he could not for the life of him think of what to say about this well-known dignitary. Dr. Buckland happily came to his rescue, and, making a funnel of his hands, whispered, " Say —there is so much to say on the subject that you don't know what to say first." Mr. Heathcote took the hint, and that utterance gave time for the particular compliment to Dr. Ingram as an author to come to his mind, and the whole party cheered the allusion to Dr. Ingram's achievement.

CHAPTER VIII.

THE BRIDGEWATER TREATISE ON "GEOLOGY AND MINERALOGY CONSIDERED WITH REFERENCE TO NATURAL THEOLOGY."

> "We can read Bethel on a pile of stones,
>   And, seeing where God *has* been, trust in Him."
>   LOWELL, *Cathedral.*

IN 1830 Dr. Buckland was requested by the trustees under the will of the late Earl of Bridgewater to write one of the eight treatises designed in accordance with the will to "justify the ways of God to man." "Geology and Mineralogy Considered with Reference to Natural Theology" is the title of the book which is best known to the public in connection with Dr. Buckland's name. That some portions of this valuable work have grown obsolete by the progress of these sciences is a matter of course. But the main argument is as powerful as ever, and has been accepted by men who, like Professor Owen and Professor Phillips, occupy an unassailable position in the scientific world ; and so little has it been superseded by any other work on the subject that Professor Boyd Dawkins uses it as a book of reference at the present day in Owens College, Manchester. The Treatise, which was six years in writing and was not published until 1836, was widely read and won a distinguished reputation for its author. Most of his work was done at night, and his habits in this respect made it difficult for him to write except at home. " I have about as much command of time here," he writes to Murchison when on a visit the year that the Treatise was published, "as a turnpike man, and as I have not your valuable military talent of early rising I cannot steal a march upon the enemy by getting over the ground before breakfast."

The third edition, brought out after the Dean's death in 1856, was edited by Professor Phillips, and prefaced by a short memoir by Frank Buckland, who thus writes of Mrs. Buckland :—

" Not only was she a pious, amiable, and excellent help-mate to my father ; but being naturally endowed with great mental powers, habits of perseverance and order, tempered by excellent judgment, she materially assisted her husband in his literary labours, and often gave to them a polish which added not a little to their merits. During the long period that Dr. Buckland was engaged in writing the Bridge-water Treatise, my mother sat up night after night, for weeks and months consecutively, writing to my father's dictation ; and this, often till the sun's rays, shining through the shutters at early morn, warned the husband to cease from thinking, and the wife to rest her weary hand."

The labour of preparing a work which broke new ground in so many directions was enormous. Speaking of the book before the British Association at Bristol in 1836, Buckland explains the causes which had delayed its appearance.

" Let any person," he says, "the least conversant with books of a similar description ; let any person who knows

13

what it is to have drawings, many of them from microscopic objects, made by artists, of new and unfamiliar subjects— let him consider that five or six different artists have been employed—that all their errors had severally to be corrected, that these engravings consist of seven hundred and five figures—then I repeat that he alone who has had a full experience of the difficulty will be able to appreciate the causes of the delay. For my own part I am astonished it has been finished so soon ; and of this I assure you, that such is the intricacy of the subject, such the tiresomeness of the details, that were the work to be done over again, no power on earth should induce me to undertake it."

In the second chapter of the Bridgewater Treatise Dr. Buckland uses an argument, which is now familiar, but was then comparatively ignored. He urged that the Bible was not written to teach scientific truth, but to reveal God and to instruct us in the Divine Life. Nay more ; he does not hesitate to say that, if the Bible had been made an adequate text-book of science, men would have found it a source of perplexity and not of enlightenment.

" We may fairly ask of those persons who consider physical science a fit subject for revelation, what point they can imagine short of a communication of Omniscience at which such a revelation might have stopped, without imperfections of omission, less in degree, but similar in kind, to that which they impute to the existing narrative of Moses ? A revelation of so much only of astronomy as was known to Copernicus, would have seemed imperfect after the discoveries of Newton, and a revelation of the science of Newton would have appeared defective to La Place ; a revelation of all the chemical knowledge of the eighteenth century would have been as deficient in comparison with the information of the present day, as what is now known in this science will probably appear before the termination of another age. In the whole circle of sciences there is not one to which this argument may not be extended, until we should require from revelation a full development of all the mysterious agencies that uphold the mechanism of the material world."

This argument is unanswerable, and in an article on the work in the *Quarterly Review* (April 1836), it is so treated by the reviewer, who regards the Treatise as one calculated to " astonish and delight all lovers of science, if any such there be who may be ignorant of the extent of the field which geology has laid open." In concluding his notice of this " most instructive and interesting volume, of which every page is pregnant with facts inestimably precious to the natural theologian," the author of the article thanks Dr. Buckland for his

" industry and research, and for the commanding eloquence with which he has called forth the very stocks and stones that have been buried for countless ages in the deep recesses of the earth to proclaim the universal agency throughout all time of one all-directing, all-pervading mind, and to swell the chorus in which all creation 'hymns His praise,' and be a witness to His unlimited power, wisdom, and benevolence."

It is interesting to observe that the *Edinburgh Review* (April 1837), in an elaborate article on geological science suggested by the same work, writes in a similar strain, and praises the book "as pregnant with the deepest instruction and calculated to inspire the most affectionate veneration for that Great Being who has made even the convulsions of the material world subservient to the civilisation and happiness of His creatures." Later on, the writer,

like his predecessor in the *Quarterly*, praises Dr. Buckland's
" lofty and impressive eloquence," and adds : " We have
ourselves never perused a work more truly fascinating, or
more deeply calculated to leave abiding impressions on the
heart."

Other criticisms were not so favourable. In one of the
few letters which have been preserved of Mrs. Buckland's,
she alludes to the attacks that were made upon her husband
in the press.

" A note from Dr. Shuttleworth thanks you for your
present (of the Bridgewater Treatise), which he considers
' to be a valuable addition to the philosophical literature
of our country, or rather of *our planet*, as we nowadays
express ourselves.' I could not resist opening this as
well as Drs. Frowd's and Simmonds' notes, and I hope
you will keep your letters of thanks for my perusal.
Keep the *St. James's Chronicles*, every one of which has a
rap at you ; but I beseech you not to lower your dignity
by noticing newspaper statements. I have not seen the
Standard nor *John Bull* ; but I hear they are in the same
strain."

The novelty of many of the conclusions at which Buckland arrived easily accounts for divergencies of opinion as
to the work. But no one who reads the Treatise can fail
to be struck with the lucidity of the style. Not even Paley
in his " Natural Theology " is clearer than Dr. Buckland,
and the second volume, which consists of plates, makes
the whole subject intelligible to persons who have never
had a scientific education. In the chapter on the Fossil
Vertebrated Animals occurs a passage upon Cuvier, which
may be quoted as a noble tribute from a distinguished
man of science to the genius of the great Frenchman.

"The result of his researches," Dr. Buckland writes, "as recorded in the 'Ossemens Fossiles,' has been to show that all fossil quadrupeds, however differing in generic, or specific details, are uniformly constructed on the same general plan, and systematic basis of organisation, as living species; and that throughout the various adaptations of a common type to peculiar functions, under different conditions of the earth, there prevails such universal conformity of design, that we cannot rise from the perusal of these inestimable volumes without a strong conviction of the agency of one vast and mighty intelligence, ever directing the entire fabric, both of past and present systems of creation. Nothing can exceed the accuracy of the severe and logical demonstrations, that fill these volumes with proofs of wise design, in the constant relation of the parts of animals to one another, and to the general functions of the whole body. Nothing can surpass the perfection of his reasoning, in pointing out the beautiful contrivances, which are provided in almost endless variety, to fit every living creature to its own peculiar state and mode of life." [1]

Of Buckland, as of Cuvier, it may be truly said that both men wrote and studied with the same high object in view, and that both, in the course and by the means of their studies, were alike impressed with an assurance of the existence of one Supreme Creator of all things, and, to quote the words of Boyle, with "the high veneration man's intellect owes to God."

The following extracts from the "Essay," as Buckland modestly called it, serve to illustrate the general argument of the whole, and the special examples by which the argument was enforced. The section which is devoted

---

[1] Bridgewater, vol. i., p. 141.

to recent discoveries earned for the author the apt title of
" the Æsop of extinct animals." [1]

On the general history of fossil organic remains
Buckland writes :—

" As 'the variety and formation of God's creatures in
the animal, vegetable, and mineral kingdoms' are specially
marked out by the founder of this Treatise as the subjects
from which he desires that proofs should be sought of
the power, wisdom, and goodness of the Creator ; I shall
enter at greater length into the evidences of this kind,
afforded by fossil organic remains, than I might have done,
without such specific directions respecting the source from
which my arguments are to be derived. . . .

"From the high preservation in which we find the
remains of animals and vegetables of each geological
formation, and the exquisite mechanism which appears in
many fossil fragments of their organisation, we may collect
an infinity of arguments, to show that the creatures from
which all these are derived were constructed with a view
to the varying conditions of the surface of the earth, and to
its gradually increasing capabilities of sustaining more com-
plex forms of organic life, advancing through successive
stages of perfection. Few facts are more remarkable in the
history of the progress of human discovery, than that it
should have been reserved almost entirely for the researches

---

[1] Mr. Waterhouse Hawkins, after a continuous mental and bodily
labour of more than three years, presented to the public notice in
the gardens of the Crystal Palace at Sydenham restorations of no less
than thirty-three extinct animals, known to us only by their fossil
remains. He told Mr. Frank Buckland that in modelling his restorations he had received the greatest assistance from the Plates in the
Bridgewater Treatise, many of which were drawn by Mrs. Buckland.
Mr. W. Hawkins gave to Mrs. Buckland the original sketch from his
own pencil, which is here reproduced, of his marvellous models of
ancient marine saurians, the originals of which are now at Sydenham.
For description see Bridgewater, p. 38.

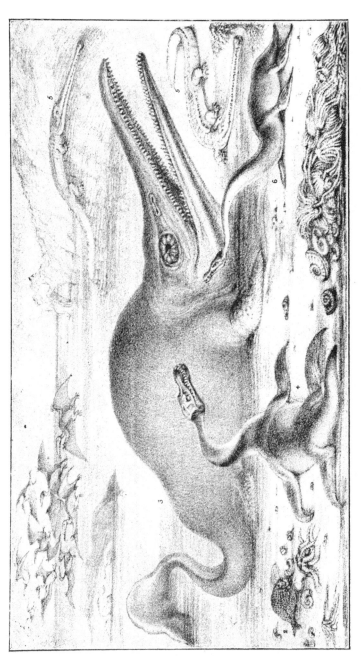

RESTORATION OF SAURIANS AND OTHER EXTINCT ANIMALS.

1. Pterodactyle.
2. Dapedius.

3. Ichthyosaurus communis.
4. Plesiosaurus macrocephalus.

5. Teleosaurus.
6. Plesiosaurus dolichodeirus.

[From a drawing by Waterhouse Hawkins.
[Face p. 198.

of the present generation to arrive at any certain knowledge
of the existence of the numerous extinct races of animals,
which occupied the surface of our planet in ages preceding
the creation of man. . . .

"We can hardly imagine any stronger proof of the unity
of design and harmony of organisations that have ever
pervaded all animated nature, than we find in the fact
established by Cuvier, that from the character of a single
limb, and even of a single tooth or bone, the form and proportions of the other bones, and condition of the entire
animal, may be inferred. This law prevails no less
universally throughout the existing kingdoms of animated
races, than in those various races of extinct creatures that
have preceded the present tenants of our planet; hence,
not only the framework of the fossil skeleton of an extinct
animal, but also the character of the muscles by which each
bone was moved, the external form and figure of the body,
the food, and habits, and haunts, and mode of life of creatures that ceased to exist before the creation of the human
race, can with a high degree of probability be ascertained.
The study of organic remains, indeed, forms the peculiar
feature and basis of modern geology, and is the main cause
of the progress this science has made since the commencement of the present century. We find certain families of
organic remains pervading strata of every age, under nearly
the same generic forms which they present among existing
organisations. Other families, both of animals and vegetables, are limited to particular formations, there being
certain points where entire groups ceased to exist and were
replaced by others of a different character. The changes
of genera and species are still more frequent; hence it has
been well observed, that to attempt an investigation of the
structure and revolutions of the earth, without applying
minute attention to the evidences afforded by organic
remains, would be no less absurd than to undertake to
write the history of any ancient people, without reference to
the documents afforded by their medals and inscriptions,
their monuments, and the ruins of their cities and temples.
The secrets of Nature, that are revealed to us by the history

of fossil organic remains, form perhaps the most striking results at which we arrive from the study of geology. It must appear almost incredible to those who have not minutely attended to natural phenomena, that the microscopic examination of a mass of rude and lifeless limestone should often disclose the curious fact, that large proportions of its substance have once formed parts of living bodies. It is surprising to consider that the walls of our houses are sometimes composed of little else than comminuted shells, that were once the domicile of other animals, at the bottom of ancient seas and lakes. It is marvellous that mankind should have gone on for so many centuries in ignorance of the fact, which is now so fully demonstrated, that no small part of the present surface of the earth is derived from the remains of animals that constituted the population of ancient seas. Many extensive plains and massive mountains form, as it were, the great charnel-houses of preceding generations, in which the petrified exuviæ of extinct races of animals and vegetables are piled into stupendous monuments of the operations of life and death, during almost immeasurable periods of past time. 'At the sight of a spectacle,' says Cuvier, ' so imposing, so terrible, as that of the wreck of animal life, forming almost the entire soil on which we tread, it is difficult to restrain the imagination from hazarding some conjectures as to the causes by which such great effects have been produced.'[1] The deeper we descend into the strata of the earth, the higher do we ascend into the archæological history of past ages of creation. We find successive stages marked by varying forms of animal and vegetable life, and these generally differ more and more widely from existing species as we go further downwards into the receptacles of the wreck of more ancient creations.

"When we discover a constant and regular assemblage of organic remains, commencing with one series of strata, and ending with another, which contains a different assemblage, we have herein the surest grounds whereon

---

[1] Cuvier, "Rapport sur le Progrès des Sciences Naturelles," p. 179.

to establish those divisions which are called geological formations. . . . Thus it appears, that the more perfect forms of
animals become gradually more abundant, as we advance
from the older into the newer series of depositions, whilst
the more simple orders, though often changed in genus
and species, and sometimes losing whole families, which
are replaced by new ones, have pervaded the entire range
of fossiliferous formations. . . . Minute examination discloses
occasionally prodigious accumulations of microscopic shells,
that surprise us no less by their abundance than their
extreme minuteness ; the mode in which they are sometimes crowded together, may be estimated from the fact
that Soldani collected from less than an ounce and a half
of stone found in the hills of Casciana, in Tuscany, ten
thousand four hundred and fifty-four microscopic cham-
bered shells. The rest of the stone was composed of
fragments of shells, of minute spines of Echini, and of
a sparry calcareous matter. Of several species of these
shells, four or five hundred weigh but a single grain ;.
of one species he calculates that a thousand individuals
would scarcely weigh one grain. . . . Similar accumulations
of microscopic shells have been observed also in various
sedimentary deposits of fresh-water formation. A striking example of this kind is found in the abundant
diffusion of the remains of a microscopic crustaceous
animal of the genus Cypris. Animals of this genus are
enclosed within two flat valves, like those of a bivalve
shell, now inhabiting the waters of lakes and marshes.
Certain clay beds of the Wealden formation below the
chalk are so abundantly charged with microscopic shells
of the Cypris Faba, that the surfaces of many laminæ into
which this clay is easily divided, are often entirely covered
with them as with small seeds. The same shells occur
also in the Hastings sand and sandstone, in the Sussex
marble, and in the Purbeck limestone, all of which were
deposited during the same geological epoch in an ancient
lake or estuary, wherein strata of this formation have
been accumulated to the thickness of nearly a thousand
feet. . . . In the case of deposits formed in estuaries, the

admixture and alternation of the remains of fluviatile and lacustrine shells with marine exuviæ, indicate conditions analogous to those under which we observe the inhabitants both of the sea and rivers existing together in brackish water near the deltas of the Nile, and other great rivers. Thus, we find a stratum of oyster shells, that indicates either the presence of salt or brackish water, interposed between limestone strata filled with freshwater shells, among the Purbeck formations ; so also in the sand and clays of the Wealden formation of Tilgate Forest we have freshwater and lacustrine shells intermixed with remains of large terrestrial reptiles, *e.g.*, Megalosaurus, Iguanodon, and Hylæosaurus ; with these we find also the bones of the marine reptiles Plesiosaurus ; and from this admixture we infer that the former were drifted from the land into an estuary, which the Plesiosaurus also having entered from the sea, left its bones in this common receptacle of the animal and mineral exuviæ of some not far distant land."

It will be remembered that in 1820 the Professor had visited the caves of Monte Bolca, when his friend Count Breüner depicted him in his "fishing" costume. It may, therefore, not be uninteresting to hear the account which he gives in the Treatise of the quarries :—

"The circumstances under which the fossil fish are found at Monte Bolca seem to indicate that they perished suddenly on arriving at a part of the then existing seas, which was rendered noxious by the volcanic agency of which the adjacent basaltic rocks afford abundant evidence. The skeletons of these fish lie parallel to the laminæ of the strata of the calcareous slate ; they are always entire, and so closely packed on one another that many individuals are often contained in a single block. The thousands of specimens which are dispersed over the cabinets of Europe have nearly all been taken from one quarry. All these fish must have died suddenly on this fatal spot, and have been speedily buried in the calcareous sediment then in the course of deposition. From the fact that certain individuals have even preserved traces of colour upon their skin, we are certain that they were entombed before decomposition of their soft parts had taken place." [1]

The Stonesfield quarry, near Oxford, which yielded such prolific spoils to his geological hammer, is described in the following words :—

"At this place, a single bed of calcareous and sandy slate, not six feet thick, contains an admixture of terrestrial animals and plants with shells that are decidedly marine ; the bones of Divelphys (Opossum), Megalosaurus, and Pterodactyle are so mixed with Ammonites, Nautili, and Belemnites, and many other species of marine shells, that there can be little doubt that this formation was deposited at the bottom of a sea, not far distant from some ancient shore. We may account for the presence of remains of terrestrial animals in such a situation by supposing their carcases to have been floated from land at no great distance from their place of sub-marine interment." [2]

It was in these Stonesfield quarries that the megalosaurus was discovered ; but, before giving Buckland's account of the monster, it may be convenient to mention the specimen which Frank Buckland calls, in his "Curiosities of Natural History," "the great gem of the Stonesfield fossils, the jaw of the Phascolotherium,[3] a small marsupial or pouched animal (hence such a big name for such a little creature : *phascolos*, a leathern bag,

---

[1] Bridgewater, vol. i., p. 124.

[2] Bridgewater, vol. i., p. 122.

[3] Lower jaw and teeth of Phascolotherium Bucklandi from great oolite (Stonesfield). Oxford Museum, table-case 14 ; Fig. 97, B.M.N.H.

and *therium*, a beast), the first, and, at one time, the sole evidence of mammalian life having existed at the earlier period of the earth's history; it has found a good, and, we trust, a lasting home in the museum at Oxford, but a few miles from the place where, ages and ages ago, it roamed over the neighbourhood of Woodstock. Little did this tiny beast think that one day its under jaw would cause Dons to open their eyes with astonishment, and Professors to tax their memories and brains for appropriate words wherewith to descant upon its beauty, and upon the deductions logically to be inferred from it as to the climate and state of animal and vegetable life at the time it existed."[1]

Cuvier,[2] to whom Dr. Buckland had dedicated his memoir on the megalosaurus (big lizard), speaks of his discovery in the following terms :—

"L'un des hommes qui honorent la géologie par les observations précises et suivies, et par la résistance la plus constante aux hypothèses hasardées, Monsieur le professeur Buckland, a fait depuis plusieurs années cette belle découverte, et j'en ai vu les pièces chez lui à Oxford en 1818 ; j'y en ai même dessiné quelques-unes ; mais il a eu, depuis, la complaisance de m'adresser le mémoire qu'il va donner sur ce sujet dans le Recueil de la Société Géologique de Londres, où il fait connaître exactement les os qu'il possède et les circonstances de leur gisement ; c'est de cet écrit que je tire les principaux matériaux du présent article."

"Although," says Buckland, "no skeleton has been found entire, so many perfect bones and teeth have been discovered in the same quarries, that we are nearly as well acquainted with the form and dimensions of its limbs, as if they had been found together in a single block of stone. From the size and proportions of these bones, as compared with existing lizards, Cuvier concludes the Megalosaurus to

---

[1] "Curiosities of Natural History," 2nd series.
[2] "Ossemens Fossiles," vol. v., p. 2. (Paris, 1825.)

have been an enormous reptile, measuring from forty to fifty feet in length, and partaking of the structure of the Crocodile and the Monitor. As the femur and tibia measure nearly three feet each, the entire hind leg must have attained a length of nearly two yards. The bones of the thigh and leg are not solid at the centre, as in crocodiles and other aquatic quadrupeds, but have large medullary cavities, like the bones of terrestrial animals. We learn from this circumstance, added to the character of the foot, that the Megalosaurus lived chiefly upon the land. . . . The form of the teeth shows the Megalosaurus to have been in a high degree carnivorous : it probably fed on smaller reptiles, such as crocodiles and tortoises, whose remains abound in the same strata with its bones. It may also have taken to the water in pursuit of Plesiosauri and fishes. The most important part[1] of the Megalosaurus yet found consists of a fragment of the lower jaw, containing many teeth. The form of this jaw shows that the head was terminated by a straight and narrow snout. . . . In the structure of these teeth we find a combination of mechanical contrivances analogous to those which are adopted in the construction of the knife, the sabre, and the saw. When first protruded above the gum the apex of each tooth presented a double cutting edge of serrated enamel. In this stage, its position and line of action were nearly vertical, and its form like that of the two-edged point of a sabre, cutting equally on each side. As the tooth advanced in growth, it became curved backwards in the form of a pruning knife, and the edge of serrated enamel was continued downwards to the base of the inner and cutting side of the tooth. . . . In a tooth thus formed for cutting along its concave edge each movement of the jaw combined the power of the knife and saw ; whilst the apex, in making the first incisions, acted like the two-edged point of a sabre. The backward curvature of the full-

---

[1] This " most important part " is in a case in an upper gallery of the Oxford Museum, while the rest of the specimen is in a separate case on the ground floor.

grown teeth enables them to retain, like barbs, the prey which they had penetrated. In these adaptations, we see contrivances, which human ingenuity has also adopted, in the preparation of various instruments of art.

"In a former chapter I endeavoured to show that the establishment of carnivorous races throughout the animal kingdom tends materially to diminish the aggregate amount of animal suffering. The provision of teeth and jaws, adapted to effect the work of death most speedily, is highly subsidiary to the accomplishment of this desirable end. We act ourselves on this conviction, under the impulse of pure humanity, when we provide the most efficient instruments to produce the instantaneous and most easy death of the innumerable animals that are daily slaughtered for the supply of human food." [1]

Those readers who are curious to see the big wild-beast and the big lizard—the Megatherium and the Megalosaurus—may see Mr. Waterhouse Hawkins's wonderful restorations of these and other fossil monsters in the lower lake of the grounds of the Crystal Palace. On a prominent point of the lake are placed some half-bird, half-bat-like creatures called Pterodactyles, which have also been discovered, together with the little opossum and the big lizard, in the Stonesfield quarries. [2]

"The structure of these animals," says Buckland, "is so exceedingly anomalous, that the first discovered Pterodactyle (or Flying Lizard) was considered by one naturalist to be a Bird, by another as a species of Bat, and by a

---

[1] Bridgewater, vol. i., p. 227.

[2] What a picture we might have of Old World life at Stonesfield if a representation could be made in the Oxford Museum of the fauna and flora found there, and of which no entire record has ever been made!

third as a flying Reptile. This extraordinary discordance
of opinion respecting a creature whose skeleton was almost
entire, arose from the presence of characters apparently
belonging to each of the three classes to which it was
referred ; the form of its head and length of neck resembling that of Birds, its wings approaching to the proportions and form of those of Bats, and the body and
tail approximating to those of ordinary Mammalia. These
characters, connected with a small skull, as is usual among
reptiles, and a beak furnished with not less than sixty
pointed teeth, presented a combination of apparent anomalies
which it was reserved for the genius of Cuvier to reconcile.
In his hands, this apparently monstrous production of the
ancient world has been converted into one of the most
beautiful examples yet afforded by comparative anatomy,
of the harmony that pervades all nature, in the adaptation
of the same parts of the animal frame to infinitely varied
conditions of existence. . . .

"The Pterodactyles are ranked by Cuvier among the most
extraordinary of all the extinct animals that have come
under his consideration. 'Ce sont incontestablement de
tous les êtres dont ce livre nous révèle l'ancienne existence,
les plus extraordinaires, et ceux qui, si on les voyait vivans,
paroîtraient les plus étrangers à toute la nature actuelle'
(Cuvier, 'Ossemens Fossiles,' vol. v.). We are already
acquainted with eight species of this genus, varying from
the size of a snipe to that of a cormorant.[1]

"In external form, these animals somewhat resembled
our modern bats and vampires : most of them had the
nose elongated, like the snout of a crocodile, and armed
with conical teeth. Their eyes were of enormous size,
apparently enabling them to fly by night. From their
wings projected fingers, terminated by long hooks, like the
curved claw on the thumb of the bat. These must have
formed a powerful paw, wherewith the animal was enabled

---

[1] Some fragments of Pterodactyle bones from the green sand,
Cambridge, must have belonged to one of gigantic dimensions, and could
not have been of less expanse from wing to wing than 27 feet.

to creep or climb, or suspend itself from trees. It is probable also that the Pterodactyles had the power of swimming, which is so common in reptiles, and which is now possessed by the vampire bat of the island of Bonin.

"Thus, like Milton's fiend, all qualified for all services and all elements, the creature was a fit companion for the kindred reptiles that swarmed in the seas, or crawled on the shores of a turbulent planet.

' The fiend,
O'er bog or steep, through strait, rough, dense, or rare,
With head, hands, wings, or feet pursues his way,
And swims, or sinks, or wades, or creeps, or flies.'
Paradise Lost, Book II., line 947

With flocks of such-like creatures flying in the air, and shoals of no less monstrous ichthyosauri and plesiosauri swarming in the ocean, and gigantic crocodiles and tortoises crawling on the shores of the primæval lakes and rivers, air, sea, and land must have been strangely tenanted in these early periods of our infant world." [1]

"As the most obvious feature of these fossil reptiles is the presence of organs of flight, it is natural to look for the peculiarities of the Bird or Bat, in the structure of their component bones. All attempts, however, to identify them with birds are stopped at once by the fact of their having teeth in the beak, resembling those of reptiles : the form of a single bone, the *os quadratum*, enabled Cuvier to pronounce at once that the creature was a Lizard : but a Lizard possessing wings exists not in the present creation, and is to be found only among the Dragons of romance and heraldry ; while a moment's comparison of the head and teeth with those of Bats shows that the fossil animals in question cannot be referred to that family of flying mammalia. As an insulated fact, it may seem to be of little moment whether a living Lizard or a fossil Pterodactyle might have four or five joints in its fourth finger,

[1] Geological Trans. (London), N. S., vol. iii., Part I.

or its fourth toe; but those who have patience to examine the minutiæ of this structure, will find in it an exemplification of the general principle, that things apparently minute and trifling in themselves, may acquire importance, when viewed in connection with others, which taken singly appear equally insignificant. Minutiæ of this kind, viewed in their cogent relations to the parts and proportions of other animals, may illustrate points of high importance in physiology, and thereby become connected with the still higher considerations of natural theology. If we examine the forefoot of the existing Lizards, we find the number of joints regularly increased by the addition of one, as we proceed from the first finger, or thumb, which has two joints, to the third, in which there are four; this is precisely the numerical arrangement which takes place in the three first fingers of the hand of the Pterodactyle. Thus far the three first fingers of the fossil reptile agree in structure with those of the forefoot of living Lizards; but as the hand of the Pterodactyle was to be converted into an organ of flight, the joints of the fourth or fifth finger were lengthened to become expansors of a membranous wing. As the bones in the wing of the Pterodactyle thus agree in number and proportion with those in the forefoot of the lizard, so do they differ entirely from the arrangement of the bones which form the expansors of the wing of the bat. The total number of toes in the Pterodactyles is usually four; the exterior, or little toe, being deficient: if we compare the number and proportion of the joints in these four toes with those of Lizards, we find the agreement as to number, to be not less perfect than it is in the fingers; we have, in each case, two joints in the first or great toe, three in the second, four in the third, and five in the fourth. As to proportion also, the penultimate joint is always the longest, and the antepenultimate, or last but two, the shortest; these relative proportions are also precisely the same, as in the feet of lizards. The apparent use of this disposition of the shortest joints in the middle of the toes of the Lizards, is to give greater power of flexion for bending round, and laying fast hold on twigs and branches of trees

14

of various dimensions, or on inequalities of the surface of the ground or rocks, in the act of climbing or running. All these coincidences of number and proportion can only have originated in a premeditated adaptation of each part to its peculiar office; they teach us to arrange an extinct animal under an existing family of reptiles; and when we find so many other peculiarities of this tribe in almost every bone of the skeleton of the Pterodactyle, with such modifications, and such only as were necessary to fit it for the purposes of flight, we perceive unity of design pervading every part, and adapting to motion in the air organs which in other genera are calculated for progression on the ground, or in the water. . . .

"With regard to their food, it has been conjectured by Cuvier that they fed on insects, and from the magnitude of their eyes that they may also have been noctivagous (wandering by night). The presence of large fossil Libellulæ, or Dragon-flies, and many other insects, in the same lithographic quarries with the Pterodactyles at Solenhofen, and of the wings of coleopterous insects, mixed with bones of Pterodactyles in the oolitic slate of Stonesfield, near Oxford, proves that large insects existed at the same time with them, and may have contributed to their supply of food. We know that many of the smaller lizards of existing species are insectivorous; some are also carnivorous, and others omnivorous; but the head and teeth of two species of Pterodactyle are so much larger and stronger than is necessary for the capture of insects, that the larger species of them may possibly have fed on fishes, darting upon them from the air after the manner of Sea Swallows and Solan Geese. The enormous size and strength of the head and teeth of the *P. crassirostris* would not only have enabled it to catch fish, but also to kill and devour the few small marsupial mammalia which then existed upon the land.

"The entire range of ancient anatomy affords few more striking examples of the uniformity of the laws which connect the extinct animals of the fossil creation with existing organised beings, than those we have been examining in the case of the Pterodactyle. We find the details of parts which, from their minuteness, should seem insignificant, acquiring great importance in such an investigation as we are now conducting ; they show, not less distinctly than the colossal limbs of the most gigantic quadrupeds, a numerical coincidence and a concurrence of proportions which it seems impossible to refer to the effect of accident, and which point out unity of purpose, and deliberate design, in some intelligent First Cause, from which they were all derived. We have seen that, whilst all the laws of existing organisation in the order of Lizards are rigidly maintained in the Pterodactyles ; still, as Lizards modified to move like Birds and Bats in the air, they received, in each part of their frame, a perfect adaptation to their state. We have dwelt more at length on the minutiæ of their mechanism, because they convey us back into ages so exceedingly remote, and show that even in those distant eras, the same care of a common Creator, which we witness in the mechanism of our own bodies, and those of the myriads of inferior creatures that move around us, was extended to the structure of creatures that, at first sight, seem made up only of monstrosities." [1]

Among the treasures of the Bucklandean Collection at Oxford is a cast of the *Mosasaurus*. It bears an inscription on the edge of it, "Given by the Museum of Natural History at Paris to Dr. Buckland," and was presented to Buckland by Cuvier, who was then omnipotent at the French Museum, as an evidence of his friendship, and of the high esteem with which he regarded him. This relic possesses so curious a history, that it may be interesting to make a further extract from the Bridgewater Treatise concerning the history of the great animal of Maestricht.

---

[1] Bridgewater, vol. i., pp. 216-227.

" The Mosasaurus (the river Meuse and *saurus*, a lizard) has been long known by the name of the great animal of Maestricht, occurring near that city in the calcareous freestone. . . . A nearly perfect head of this animal was discovered (1780) in the quarries under the hill of St. Pierre, by Dr. Hofmann, Surgeon to the Forces quartered in the town of Maestricht.[1]   This celebrated head during many years baffled all the skill of Naturalists : some considered it to be that of a whale, others of a crocodile ; but its true place in the animal kingdom was first suggested by Adrian Camper, and at length confirmed by Cuvier. By their investigations it is proved to have been a gigantic marine reptile, most nearly allied to the Monitor.[2]   The geological epoch at which the *Mosasaurus* first appeared seems to have been the last of the long series during which the oolitic and cretaceous groups were in process of formation.   In these periods the inhabitants of our planet seem to have been principally marine, and some of the largest creatures were saurians of gigantic stature, many of them living in the sea, and controlling the excessive

[1] It is recorded of this fossil that one of the canons of the cathedral church of Maestricht brought an action at law against the discoverer, Dr. Hofmann, and obtained possession from him ; but he was not allowed to hold his prize long, for, when the French Revolution broke out, and the armies of the Republic advanced to the gates of Maestricht, 1795, the town was bombarded, and at the suggestion of the committee of *savants* who accompanied the French troops to select their share of the plunder, the artillery was not allowed to play on that part of the town in which the celebrated fossil was known to be preserved.   After the capitulation of the town, it was seized and carried off in triumph. The specimen has since remained in the museum of the Jardin des Plantes in Paris, and a cast of it is now in the British Museum.

[2] The monitors form a genus of lizards, frequenting marshes and the banks of rivers in hot climates ; they have received this name from the prevailing but absurd notion that they give warning by a whistling noise of the approach of Crocodiles and Caymans.   One species the Lacerta Nilotica, or lizard of the Nile, which devours the eggs of Crocodiles, has been sculptured on the monuments of ancient Egypt.

increase of the then existing tribes of fishes. From the
lias upwards, to the commencement of the chalk formation,
the Ichthyosauri (*ichthus*, a fish) and Plesiosauri (*plesios*,
near to) were the tyrants of the ocean; and just at the
point of time when their existence terminated during the
deposition of the chalk, the new genus Mosasaurus appears
to have been introduced, to supply for a while their
place and office,[1] being itself destined in its turn to give
place to the Cetacea of the tertiary periods. As no saurians
of the present world are inhabitants of the sea, and the
most powerful living representatives of this order, viz. the
crocodiles, though living chiefly in water, have recourse
to stratagem rather than speed for the capture of their
prey, it may not be unprofitable to examine the mechanical
contrivances by which a reptile, most nearly allied to the
monitor, was so constructed as to possess the power of
moving in the sea with sufficient velocity to overtake and
capture such large and powerful fishes as, from the enormous
size of its teeth and jaws, we may conclude it was intended
to devour.

"The head and teeth point out the near relations of
this animal to the monitors; and the proportions maintained throughout all the other parts of the skeleton
warrant the conclusion, that this monstrous monitor of
the ancient deep was five-and-twenty feet in length,
although the longest of its modern congeners does not
exceed five feet. The head here represented measures
four feet in length; that of the largest monitor does not
exceed five inches. The most skilful anatomist would be
at a loss to devise a series of modifications by which a
monitor could be enlarged to the length and bulk of a
grampus, and at the same time be fitted to move with
strength and rapidity through the waters of the sea; yet
in the fossil before us we shall find the genuine characters

[1] Remains of the Mosasaurus have been discovered by Dr. Mantell
in the upper chalk near Lewes, by Professor Owen in the upper chalk
of both Kent and Sussex, and by Dr. Morton in the green sand of
Virginia.

of a monitor maintained throughout the whole skeleton, with such deviations only as tended to fit the animal for its marine existence.

"The Mosasaurus had scarcely any character in common with the crocodiles, but resembled the iguanas, in having an apparatus of teeth fixed on the pterygoid bone and placed in the roof of its mouth, as in many serpents and fishes, where they act as barbs to prevent the escape of their prey. The other parts of the skeleton follow the character indicated by the head. The vertebræ are all concave in front, and convex behind; being fitted to each other by a ball and socket joint admitting easy and universal flexion. From the centre of the back to the extremity of the tail, they are destitute of articular. apophyses, which are essential to support the back of animals that move on land : in this respect they agree with the vertebræ of dolphins, and were calculated to facilitate the power of swimming ; the vertebræ of the neck allowed to that part also more flexibility than in the crocodiles. The tail was flattened on each side, but high and deep in the vertical direction, like the tail of a crocodile, forming a straight oar of immense strength to propel the body by horizontal movements analogous to those of sculling. Although the number of caudal vertebræ was nearly the same as in the monitor, the proportionate length of the tail was much diminished by the comparative shortness of the body of each vertebra ; the effect of this variation being to give strength to a shorter tail as an organ for swimming ; and a rapidity of movement which would have been unattainable by the long and slender tail of the monitor, which assists that animal in climbing. There is a further provision to give strength to the tail, by the *chevron* bones being soldered firmly to the body of each vertebra, as in fishes. The total number of vertebræ was one hundred and twenty-three, nearly the same as in the monitors, and more than double the number of those in the Crocodiles. The ribs had a single head, and were round, as in the family of lizards. Of the extremities, sufficient fragments have been found to prove that the Mosasaurus, instead of legs, had four large paddles, resembling those of the Plesiosaurus and the Whale : one great use of these was probably to assist in raising the animal to the surface, in order to breathe, as it apparently had not the horizontal tail, by means of which the Cetacea ascend for this purpose. All these characters unite to show that the Mosasaurus was adapted to live entirely in the water ; and that although it was of such vast proportions compared with the living genera of these families, it formed a link intermediate between the Monitors and the Iguanas. However strange it may appear to find its dimensions so much exceeding those of any existing Lizards, or to find marine genera in the order of Saurians, in which there exists at this time no species capable of living in the sea, it is scarcely less strange than the analogous deviations in the Megalosaurus and Iguanodon, which afford examples of still greater expansion of the type of the Monitor and Iguana, into colossal forms adapted to move upon the land. Throughout all these variations of proportion, we trace the persistence of the same laws, which regulate the formation of living genera ; and from the combinations of perfect mechanism that have, in all times, resulted from their operation, we infer the perfection of the wisdom by which all this mechanism was designed, and the immensity of the power by which it has ever been upheld." [1]

One more extract shall be given, a brief but eloquent passage on Fossil Footsteps—the marks of reptiles of whose bones no remains have been found.

" The Historian or the Antiquary may have traversed the fields of ancient or of modern battles ; and may have pursued the line of march of triumphant conquerors, whose armies trampled down the most mighty kingdoms of the world. The winds and storms have utterly obliterated the ephemeral impressions of their course. Not a track remains of a single foot, or a single hoof, of all the countless millions

---

[1] Bridgewater, vol. i., chap. xiv.

of men and beasts whose progress spread desolation over
the earth. But the Reptiles that crawled upon the half-
finished surface of our infant planet, have left memorials
of their passage, enduring and indelible. No history has
recorded their creation or destruction ; their very bones are
found no more among the fossil relics of a former world.
Centuries, and thousands of years, may have rolled away,
between the time in which these footsteps were impressed
by Tortoises upon the sands of their native Scotland, and
the hour when they were again laid bare, and exposed to
our curious and admiring eyes. Yet we behold them
stamped upon the rock, distinct as the track of the passing
animal upon the recent snow ; as if to show that thousands
of years are but as nothing amidst Eternity—and, as it
were, in mockery of the fleeting perishable course of the
mightiest potentates among mankind." [1]

The variety and number of these impressions have created
a new science, and Ichnology has taken a definite place as
a branch of palæontological research. It may be added that
the slabs bearing the footprints to which he alludes in the
Treatise had been used to form a garden wall, from whence
four of them were taken, full of beautiful impressions of the
feet of these animals, with a cast of the nails as perfect as
if they had been taken in wax. Dr. Buckland was in the
quarry himself, and assisted one of the workmen to raise a
slab on which were these prints, which had never seen the
sun since the time they were first made. It was while
he was writing the Bridgewater that a slab of sandstone
with these footmarks had been sent him to decipher. He
was greatly puzzled ; but at last, one night, or rather
between two and three in the morning, when, according to

---

[1] Bridgewater, vol. i., chap. xiv.

his wont, he was busy writing, it suddenly occurred to him that these impressions were those of a species of tortoise. He therefore called his wife to come down and make some paste, while he went and fetched the tortoise from the garden. On his return he found the kitchen table covered with paste, upon which the tortoise was placed. The delight of this scientific couple may be imagined when they found that the footmarks of the tortoise on the paste were identical with those on the sandstone slab. Lecturing one day in Scotland on the fossil footsteps of animals, including the Cheirotherium,[1] one of his auditors at the end of the lecture referred to his diagrams exhibited, and said : " It seems, Dr. Buckland, from your drawings that all your animals walked in one direction."

" Yes," was the reply. " Cheirotherium was a Scotchman, and he always went south."

Professor Buckland finishes his book with the following words :—

" The whole course of the inquiry which we have now conducted to its close, has shown that the physical history of our globe, in which some have seen only waste, disorder, and confusion, teems with endless examples of economy, and order, and design ; and the result of all our researches, carried back through the unwritten records of past time, has been to fix more steadily our assurance of the existence of one supreme Creator of all things, to exalt more highly our conviction of the immensity of His perfections, of His might and majesty, His wisdom, and goodness, and all-sustaining providence ; and to penetrate our under-

---

[1] The form, shape, and structure of the creature who made the footprints being unknown, the name of Cheirotherium, or beast with a hand, was given to it.

standing with a profound and sensible perception of the 'high veneration man's intellect owes to God.'[1]

" The Earth from her deep foundations unites with the celestial orbs that roll through boundless space, to declare the glory and show forth the praise of their common Author and Preserver ; and the voice of Natural Religion accords harmoniously with the testimonies of Revelation, in ascribing the origin of the universe to the will of one eternal and dominant Intelligence, the Almighty Lord and supreme First Cause of all things that subsist—'the same yesterday, to-day, and for ever,' 'before the mountains were brought forth, or ever the Earth and the World were made, God from everlasting and world without end.'"[2]

---

[1] Boyle.
[2] Bridgewater, vol. i., chap. xxiv.

CHAPTER IX.

WESTMINSTER.

" I NEVER," said Sir Robert Peel, " advised an appointment of which I was more proud, or the result of which was in my opinion more satisfactory, than the nomination of Dr. Buckland to the Deanery of Westminster."

The appointment was made in 1845, in succession to Dean Wilberforce, who was promoted to the See of Oxford. Soon afterwards Dean Buckland was inducted to the living of Islip, near Oxford, bequeathed by Edward the Confessor to the Abbot of Westminster. Mrs. Buckland writes to Sir Philip Egerton from Christ Church in November 1845 :—

" It is indeed true that Dr. Buckland is to be Dean of Westminster. I have one son in the Treasury ; the other, Frank, will soon also be a resident in London, pursuing his call to surgery. To have a home for these boys would of itself be a recommendation to me for a permanent residence in London, and Sir Robert Peel's kindness has conferred upon my husband the only piece of preferment that would suit him in all respects. It comes wholly unexpected ; while we were at Havre this summer Peel offered Dr. Buckland the Deanery of Lincoln, which he declined, and never supposed Westminster would be thought of for him. I think Sir R. Peel has shown much moral courage in making choice of a person of science, for it was sure to raise a clamour, and among good people too. It has always been quite unintelligible to me how it happens that on the Continent, where there is far less religion than in England, a man who cultivates Natural History, who studies only the works of his Maker, is highly considered and raised by common consent to posts of honour, as were Cuvier, Humboldt, etc.,—while, on the contrary, in England, a man who pursues science to a religious end (even who writes a Bridgewater Treatise) is looked upon with suspicion, and, by the greatest number of those who study only the works of man, with contempt. Perhaps you can comprehend this anomaly, I cannot." [1]

She adds : " The house is large and very good, but it does not look like a very lively abode, for it opens into the Abbey and contains the Jerusalem Chamber."

The Deanery would indeed easily make four houses, the different wings being separated by large landings and passages ; and there were sixteen staircases. A long corridor, in which hung portraits of former deans, led into the Abbey, Abbot's Place or Palace, and death-chamber of

---

It is noteworthy that Professor Burdon Sanderson, in his late address at the British Association, had still cause to lament the little assistance and encouragement that scientific research receives, either from the Government or the nation, in Great Britain. Buckland in 1819, on his appointment as first Professor of Geology, in his inaugural address, when speaking of this "new learning," says: "For some years past these newly created sciences have formed a leading subject of education in most Universities on the Continent." Professor Sanderson tells us, seventy-four years afterwards, " Those who desire either to learn the methods of research or to carry out scientific inquiries have to go to Berlin, to Munich, to Breslau, or to the Pasteur Institute in Paris, to obtain what England ought long ago to have provided."

the Lancastrian King whose death it was foretold should
take place at Jerusalem.

> "Bear me to that chamber; there I'll lie,
> In that Jerusalem shall Harry die."

Dean Buckland thought that the antechamber through
which it was approached was probably the scene of Prince
Henry's wild grief over the crown. In a later day the
Jerusalem Chamber has become a household word as
the room in which meets the restored Convocation of
Canterbury.

The entrance to this oldest part of the Deanery remains
exactly as it was in 1848. In the "Robing Room," as
the antechamber was called, might be found at all seasons
of the year a blazing fire, and here for thirty years
the excellent portress, Mrs. Burrows, was in attendance
twice daily, to air the linen surplices of the canons in
residence, as it was highly necessary that these elderly
dignitaries should be protected as far as possible from the
well-known deadly cold of the Abbey.[1]

---

[1] " The apartments of the Abbot of Westminster are nearly in the
same state, at the present hour, as when they received Elizabeth (widow
of Edward IV.) and her train of young princesses. The noble stone
hall, now used as a dining-room by the students of Westminster
School, was, doubtless, the place where Elizabeth seated herself in her
despair '*alow* on the rushes, all desolate and dismayed.' Still may
be seen the circular hearth in the midst of the hall, and the remains of
a louvre in the roof, at which such portions of smoke as chose to leave
the room departed. But the merry month of May was entered when
Elizabeth took refuge there, and round about the hearth were arranged
branches and flowers, while the stone floor was strewn with green
rushes. At the end of the hall is oak panelling latticed at top, with
doors leading by winding stone stairs to the most curious nests of little

The Abbot's quarters were over the antechamber, and from them he had communication both with the Abbey and the Jerusalem Chamber. These rooms have of late years been repaired, and Dean Bradley now occupies the upper rooms. In Dr. Buckland's occupation of the Deanery, the old wainscot was so dilapidated and the rooms so cold and dismal that his servants disliked sleeping in them, and complained of the queer noises and gusts of wind blowing their candles out. At length, one stormy night, a piece of the wainscot of the narrow passage leading into the Abbot's gallery in the south-west end of the Abbey nave fell down with a crash, and discovered a well-like

rooms that the eye of antiquarian ever looked upon. These were, and still are, the private apartments of the dignitaries of the Abbey, where all offices of buttery, kitchen, and laundry are performed under many a quaint gothic arch, in some places, even at present, rich with antique corbel and foliage. This range, so interesting as a specimen of the domestic usages of the middle ages, terminates in the Abbot's own sanctum or private sitting-room, which still looks down on his lovely quiet flower garden. Nor must the passage be forgotten leading from this room to the corridor furnished with lattices, now remaining, where the Abbot might, unseen, be witness of the conduct of his monks in the great hall below. Communicating with these are the State apartments of the Royal Abbey, larger in dimensions and more costly in ornament, richly dight with painted glass and fluted oak panelling. Among these may be noted especially the organ-room, and the antechamber to the great Jerusalem Chamber—which last was the Abbot's state reception-room, and retains to this day its gothic window of painted glass of exquisite workmanship, its curious tapestry and fine original oil portrait of Richard II."—AGNES STRICKLAND'S *Queens of England*, vol. iii., p. 409.

Miss Strickland adds in a note that "the fireplace, before which Henry IV. expired, had been enriched by Henry VII. with elaborate wood entablatures, bearing his armorial devices; an addition which is the most modern part of this exquisite remnant of domestic antiquity."

opening. The alarm was great among the domestics ; but the Dean's sons were delighted at the discovery, and, having first ascertained that the air was pure by letting down a lighted candle, one of them descended by a rope and found a worm-eaten wooden bedstead and table, both in a state of crumbling decay. It was said to have been one of Dean Atterbury's hiding-places. Another of these hiding-places was in the wall of the library, a fine old room of sixty feet in length over the south cloisters. The drawing-room extends over the entrance to the Deanery from the cloisters and over the college kitchen. Under the floor of these rooms the rats had taken up their quarters, and when the house was quiet would run riot in all directions. These invisible guests, for none were ever seen, were the horror of the servants ; but the Dean, to prevent his children from being frightened, told them stories of the rats' clever doings, and how on one occasion they emptied a small cask of choice apricot wine, which his aunt had made for him in his college days, by dipping their tails into a hole that they had gnawed.

Buckland may be said to have kept open house at the Deanery. Friends were always coming to breakfast or to luncheon ; and this continuous stream of visitors served to fill his home with life and movement. The house was the centre also to which men of science resorted, and where many of their discoveries were explained or illustrated. The following note to Professor Faraday, written on June 13th, 1849, will serve as an example :—

" MY DEAR PROFESSOR,—If you can give us the pleasure of your company at lunch to-morrow at two, or any time between two and four, you will meet William Harcourt and some other naturalists, and see chloroform administered to Beast, Bird, Reptile, and Fishes.

"Very truly yours,

"W. BUCKLAND."

Among the interesting " fixtures " at the Deanery is a drawing, by Canaletti, of the procession of the Knights of the Bath, painted for Dean Wilcocks in 1747, who, like his predecessor Atterbury, also held the See of Rochester. The Dean of Westminster is *ex-officio* Chaplain of the Order, and the tradition of the picture is that Bishop Wilcocks was so proud of the position assigned him in the procession of walking next the King, that he caused the picture to be painted in order to commemorate it, and to mark as well the completion of Mr. Christopher Wren's towers. The Dean of Westminster on all official occasions wears the badge of the Order, attached to a wide red ribbon. The badge is emblematic of the sacredness of the Order— three garlands twisted together in honour of the Holy Trinity, and supposed to be derived from Arthur, founder of British chivalry. The motto is " Tria numina juncta in uno," and there is a rose, shamrock, and thistle in the centre. The Dean wears the robes on a " collar day " when he goes to Court.

The leads over the rambling old Deanery made a delight-ful playground for Buckland's children, who found that the novelty of growing mustard and cress in boxes on the roof was quite as interesting as sowing their names in the Oxford soil. There was much more light and sunshine on the leads than in the high-walled Oxford college garden, and they could always find a snug sheltered corner, whichever way the wind blew. But their favourite leads were those over the drawing-room, college hall, and Jerusalem Chamber, looking west. Magnificent sunsets were to be seen from these, particularly in the short winter days, when the wreaths of blue smoke came curling up from the chimneys of the low red-tiled roofs of old Westminster slums, and formed into fantastic-shaped purple and golden and crimson clouds as they caught the rays of the setting sun over St. James's Park and Buckingham Palace.

It was natural that the Dean, with his turn for geology and sanitary science, should carefully examine the soil on which the Abbey is built, and this is his report :—

" Thorney Island, the site of the Abbey and adjacent parts of Westminster, between the Thames and the lake in St. James's Park (which was once a swampy creek crossing between Charing Cross and Whitehall into the Thames), is a peninsula of the purest sand and gravel, which may be seen in the foundations of the Abbey and in the new deep graves in the Churchyard of St. Margaret's. The surface of the peninsula is several feet above high water mark ; its north frontier is marked by the steps ascending from the Horse Guards Parade to Duke Street, and by the Terrace, covered with houses, on the south of Birdcage Walk, whence it extends under Wellington Barracks to Buckingham Palace Gardens and Hyde Park. By the isthmus under this terrace, the peninsula of Thorney Island is connected with the gravel beds of Hyde Park, from whence the rain-water which fills the lower region of that gravel, and of the gravel in the Palace Gardens, has unbroken communication with the pure sand and gravel of the so-called Thorney Island (really a peninsula), and hence pure and much sought after water is supplied to the well and pump in Dean's Yard, and other wells in St. Peter's College, and to a pump near the north end of St. Margaret's Church."

15

No doubt this stream, which Dean Stanley calls the "vivifying centre of all that has grown up and around," had much to do with the monks' settlement on Thorney Island 1280 years ago. The monks practised the "healing art," and, though medical skill in those days was rude and simple enough, the monks knew that the secret of good health consisted in drinking pure water, and hence a Holy-well, or a Wishing-well, is to be met with in the precincts of most ruined Abbeys. No local traditions, it has been said, are "so durable as those writ in water." In 1845 the bright pure water from the old pump in Dean's Yard was still considered beneficial as an eye water, and Buckland prescribed it as such with the best results. London was not then supplied with water, as at the present time. Every day, and all the morning long, might be seen a continuous stream of water-carriers—men, women, and children—coming for the life-giving beverage. But it was in the middle of the day, when the "boys" came rushing out of school, that the scene became exciting between water-carriers and scholars. Buckets were hurled over the tall iron railings enclosing the playground, alongside which stood the famous pump; the wooden yokes and chains, upon which the buckets hung, followed; pitchers were seized, and the contents thrown in all directions. Great was the scrimmage; plentiful the splashing, and loud the cracking of pottery; boisterous often were the jokes; and lively was the merriment for a few minutes,—and then, boy-like, some other diversion was thought of, and the lads in their quaint black-tailed coats and white "chokers" dispersed.

But these merry scenes belong to the past. In making the Metropolitan Railway, twenty-four years ago, the spring, which supplied the Dean's Yard pump, and formed on its way the King's scholars' pond in Tothill Fields, was cut across ; and two engines are now employed, night and day, at the Victoria underground station, one pumping away the water from the spring rising in Hyde Park at the rate of 1200 gallons per minute, the other pumping away the sewage from the King's scholars' sewer. By draining the subsoil at Westminster, the Dean's Yard well is dried up, as also several other wells in the neighbourhood ; and the trees in the Dean's Yard are, it is to be feared, in danger of dying from drought.

As Dean of Westminster the busiest portion of Dean Buckland's always busy life began, and in all the good works which were set on foot he was warmly seconded by his wife. Yet he was never so busy as to be prevented from journeying to Oxford to lecture on his favourite science. Rising soon after seven, he worked incessantly till two or three o'clock the next morning, allowing himself scarcely time for meals, and less for recreation. One of the practical tasks to be accomplished was the removal of many great abuses that had crept into Westminster School. " In that foundation," Sir Roderick Murchison writes, " education could be no longer obtained except at costly charges, and even when these were paid, the youths were ill fed and worse lodged. All these defects were speedily rectified by the vigour and perseverance of Dean Buckland. The charges were reduced ; good diet was provided ; the rooms were well ventilated, and the buildings properly under-

drained ; so that, these physical ameliorations accompanying a really sound and good system of tuition, the fame
and credit of this venerable seminary was soon restored."

It is difficult, in the light of modern sanitary reforms, to
realise the condition of the school about fifty years ago.
Among a large collection of MS. papers in the Oxford
Museum, chiefly consisting of notes of lectures, to which
Professor Green has kindly allowed the biographer to have
access, is found a practical letter [1] of Buckland's, giving
details of his proposed alterations, and announcing a
promised subscription from Her Most Gracious Majesty
the Queen of £500, and from the Archbishop of York
£300, and further donations from old Westminsters.

Dean Buckland followed the precedent set him by Dean
Atterbury in appealing to the Crown for a subscription
towards the contemplated improvements. As in the case
of his predecessor, the domestic comfort of the Queen's
scholars was the first matter to engage his attention. In

---

[1] "*Improvements in Westminster School.*—"The Dean and Chapter
of Westminster take this method of making known to the old Westminsters that they have resolved to increase the comfort and diminish
the expense of the Queen's scholars in the following manner.

" 1st. By providing all their meals at the cost of the Establishment.

" 2nd. By fitting up large and convenient rooms for study, etc., in
the entire cloister under the dormitory.

" 3rd. By building a sanatorium at the end of the dormitory, with
rooms for a resident matron.

" 4th. By refitting the present lavatory and necessary offices with
improved hydraulic apparatus.

" 5th. By undertaking that the necessary charges on the Queen's
scholars shall not exceed £45 per annum, exclusive of books, clothes,
washing, and journeys, and the leaving fees, if the subscriptions should

many ways, but happily not in all, these two Deans resembled each other in character. Both were men of powerful intellects and of exhaustless energy ; both were eager to remove abuses and to attack prejudices ; and both possessed the gift of persuasive eloquence. Dean Buckland, however, was eminently truthful : the most splendid speech Atterbury ever delivered was in vindication of his innocence when charged with intriguing for the Pretender. Yet it is known that he had been plotting with the Jacobites all along, and on the death of Queen Anne had even offered Ormond to proclaim the Pretender at Charing Cross in his lawn sleeves.

In 1713 the School dormitory was in the monks' granary on the west side of Dean's Yard. " The gaping roof and open windows freely admitted rain and snow, wind and sun ; the beams cracked and hung with cobwebs, the cavernous walls with many a gash inflicted by youthful

---

be adequate to the costs of the contemplated improvements, which are estimated at from £3,000 to £4,000.

" 6th. It is intended in no degree to diminish the present expenses of the Dean and Chapter, and that all reductions of charges that may arise from better management shall be for the benefit of the Queen's scholars.

" The Dean and Chapter having ascertained that the present dormitory was built more than a century ago, by contributions from persons educated at Westminster, in addition to large grants from the Crown and from Parliament, have thought it reasonable to appeal again to the Crown and to old Westminsters of the present time for their aid to render more accordant with modern manners the building which has hitherto, with much inconvenience, been applied to the manifold purposes of station, study, and dormitory.

" WILLIAM BUCKLAND (Dean).
" *June 24th*, 1846."

Dukes and Earls in their boyish days ; the chairs scorched
by many a fire and engraven deep with many a famous
name."[1]   Again and again Dean Atterbury urged its
rebuilding in the college gardens ; but the Canons preferred that it should remain where it was, as their houses
looked on the gardens.   It was only by the casting vote
of the Dean that a motion was carried in favour of rebuilding the dormitory over the wide cloister which
extended along the gardens' western side.

Buckland found that Dean Atterbury's dormitory, after
over a hundred years' use as bedroom, sitting-room, and
play-room, was in a most dismal condition,—with the
walls blackened by smoke, and, here and there, hung with
moth-eaten green baize curtains ; the tables and lockers
seamed and scarred in all directions, and the floor——
Taking his children to see the place, their father asked, "Well,
children, what's this floor like ? "   The answer was prompt.
" The fossil ripple marks in our hall at home."   (A fossil
slab of ripple marks now in the Oxford Museum.)   The
floor was only cleaned once a year, so that its rough
surface was not to be wondered at, as the boys did a great
deal of cooking there amongst their other diversions.
The windows were prison-like, small and near the ceiling.
Mr. F. H. Forshall, the School chronicler, relates that, fifty
years ago, " as a rule, the windows were kept broken and
a slide was sometimes formed down college in a time of
hard frost.   On one occasion the floor was converted into
a draught board ; the Under Elections formed the pieces, and

---

[1] Dean Stanley's "Memorials of Westminster."

two seniors, standing on tables, directed their movements. When a king was made he was represented by one of the bigger boys with a small one on his back." [1]

The lavatories were in a far worse condition than those of Winchester, to which Buckland had been accustomed in his youth. A ditch filled with black mud—a creek of the Thames it was said to be—came up as far as these buildings; but apparently no tide ever succeeded in washing back into the river any of its murky contents.

Such insanitary conditions were intolerable to Buckland, and he set himself with characteristic energy to improve both the dormitory and the lavatories. His scientific reputation and his determination overpowered all resistance. Yet a weaker man would have been powerless. " I doubt," writes the Rev. E. Marshall, one of the late masters at Westminster, "if any one with a less commanding scientific reputation than Dr. Buckland, even with all the power of the Dean, could have overcome the prejudice which at that time was entertained against the alterations."

The cloister under the dormitory in the college garden was converted into day-rooms; a matron's house and sick-room were instituted; and convenient offices were built. These thorough reforms may be said to have been carried out by his force of will. Mr. Forshall states that "the advantages to the boys of these reforms were almost incalculable. Thinking that the Queen's scholars were entitled to free commons, he provided breakfasts in Hall;

---

[1] F. H. Forshall, "Westminster School Past and Present."

and by erecting a sanatorium, obviated the necessity of their using boarding houses, thus effecting a saving to each boy's parents of at least £30 per annum. Fully to appreciate this we must remember that the School had no fees of its own, but was entirely dependent on the Dean and Chapter ; so what was spent was practically taken from the incomes of the Dean and Prebendaries."

In all these new arrangements Dean Buckland took a personal interest. Every Sunday morning,—after the Abbey service was over, and after he had, according to the old Christ Church fashion, taken his children for a walk in St. James's Park to see the water-fowl, and to be rewarded with a penny if they spied any new importation among the feathered flock,—he took them the round of the School premises, beginning always with the "sick house," chatting with any of the seniors they met, inquiring how the new arrangements he had made for them suited their convenience, and asking them for practical suggestions. He also constantly visited the college kitchen to see that the food provided for the boys was of proper quality and properly dressed, and daily a "bever"[1] loaf was sent in from College Hall for his own breakfast.

Among other additions to the comfort of the boys, he secured an excellent butler, Cleghorn by name, who had been a prominent member of the police force, and almost killed by sheep stealers. Cleghorn used to tell the boys

---

[1] "Bever," from the old French *beuve = boire*, to drink ; refreshments consisting of bread and beer, formerly served in the afternoon in College Hall, answering to our five o'clock tea, now applied to the rolls on the Hall table.

of the trouble the Dean took in looking after the details of food and drink, and instead of the everlasting mutton, they had new dishes and pies and puddings never before seen at college dinners. The severe conservatism, however, of the Westminster boys at first resented the innovation, and the puddings were thrown at the cook's head! But, as Mr. Forshall, himself an ardent Old Westminster, and the chronicler of the history of his beloved School, admits, " We were all, notwithstanding, extremely glad afterwards of the improvements introduced by the Dean in our diet." And he adds : " As a Queen's scholar I have a lively recollection of the Dean's presence, and of his loving, hearty way of speaking. I very vividly remember also his introducing Dr. Liddell to us in the great schoolroom as our new Head Master, the first who had not been educated at Westminster. The Dean made a most earnest and affectionate speech to us, standing in front of the sixth form by the side of Dr. Liddell. Though his figure and manner are before my eyes at this moment, the words have vanished save, ' I present to you—Dr. Liddell—a lexicographer of European reputation.' "

The Dean often gave lectures on "common subjects " to the boys in their new sitting-room. Soon after Dr. Liddell's appointment as Head Master, Buckland gave a *conversazione* in the chamber recently constructed under the dormitory. Invitations were issued to meet the lately consecrated Bishop of Adelaide, an old Westminster scholar. Mr. Marshall thus describes the gathering :—

" The dignified proportions and solidity of the room, the crude white of the walls glaring in the light of the unshaded gas,[1] the fresh and obtrusively level floor, and the unusual sight of ladies in the boys' sitting-room, were all contrary to ordinary experiences.

"The Bishop proposed that some of the great public schools of England should contribute a sum of money to buy land, then naturally very cheap in the colony, and that the land so acquired should be the endowment of scholarships in his college, to be named respectively after the various schools. He paid his own school the compliment of coming to it first. The proposal was received with acclamation, and money afterwards subscribed to carry out the object. The proceedings of the evening concluded by the handing of some choice Lunel wine, to which no one made any conscientious objections. I vividly recall the geniality of the Dean's manner, and the kindness and hospitality shown me by both the Dean and Mrs. Buckland at the Deanery."

There is a tablet, it is said, in the Hall of St. Peter's College, Adelaide, which records the foundation of this scholarship, which took place on St. Peter's Day, 1847.

Dr. Short was consecrated Bishop of Adelaide with three other colonial bishops,—Dr. Gray, Bishop of Capetown; Dr. Perry, Bishop of Melbourne; Dr. Tyrrell, Bishop of Newcastle : the bishoprics of Adelaide and Capetown being endowed by the munificence of Miss (now Baroness) Burdett-Coutts. It was while this solemn ceremony, which lasted four hours, was going on in the Abbey that a fight took place in the "Green," the square enclosure within the cloisters. Mr. Walter Severn, the son of the well-known artist who soothed the dying hours of Keats, thus tells this highly characteristic story of Westminster School life

---

[1] Then for the first time used in college.

"fifty years since." Fights, it may be added, have since been discontinued, owing to the rule being made that they must take place before early morning school :—

"It was in 1847, a few weeks before I left Westminster, that the following incident occurred :—I was rowing up the river in one of the 'heavy fours' which went out daily during the rowing season, and as we were returning our boat had a race with one of the other 'fours,' during which a 'foul' occurred. The two boats drifted close together, and our oars got mixed up. At this moment a tall youth in the other boat snatched some of the jackets out of ours and threw them into the water. On this, my crew at once called on me, as the biggest boy in the boat, to knock him over, which I promptly did, with an oar. Immediately on landing at the 'barges,' he came up and challenged me in the usual formal style to fight in the 'Green,' and the news was quickly carried down College Street to Dean's Yard that a fight would take place next day. With thoughts of the morrow in my head, I wended my way home to James Street, Buckingham Gate, where my father then lived. Should I inform my parents about it? I had often confided in my mother, who was quite a Spartan mother, and not likely to interfere in a fair fight; but my father was essentially, as he called himself, a man of peace; and I decided that I must not let out a word at home. He had already been extremely put out on the occasion of a former fight I had with a boy called Stanton. I may mention here that fights in those days were conducted exactly like prize-fights, and were not interfered with by the authorities. I have heard it said that about 1843 some of the fights that took place in the Green actually appeared in *Bell's Life*; but this was too much even for Williamson, who interfered and stopped the reports. After partaking of late dinner or supper at home, it suddenly occurred to me that my braces were worn out and shabby, and I was determined to get new ones, so as to make a good appearance when 'stripped' for the coming fight. I had considerable difficulty in persuading my mother to give me the necessary money, as she could not understand my wishing to get them at so late an hour. A shop in York Street was still open, and I secured a pair of bright-red braces, which were such a novelty that there was slight applause from the ring on the morrow when my outer garments were removed. I may as well mention for the uninitiated that this ancient fighting-green is the quiet, peaceful-looking grass-plot in the centre of the cloisters, under the shadow of our grand and venerable Abbey. The Green presented an animated appearance, with an unusually large ring, which took up most of the space. At that time the cloisters all round were very much out of repair, almost in ruins, and on two sides the broken arches extended to the ground, so that there were many exits to and from the Green.

"The day was one of the many 'saints' days' which were kept as holidays. I think Dryden, who was an 'Old Westminster,' alludes to the extraordinary number of these holidays in his time.

"There happened to be a grand consecration of four colonial bishops in the Abbey, so that we were not without solemn music to give *éclat* to our little entertainment outside. I distinctly remember that I went into this fight with a cheerful heart and a perfectly clear conscience. My antagonist was not a popular boy, and the fact that I was going to fight him was very much approved of. He was bigger and stronger than I was, but I was more active and a better boxer, having practised the art with a prize-fighter who used to give lessons to some of the older boys.

"Round succeeded round for more than an hour, until we were both becoming somewhat exhausted, when a sudden interference took place which stopped the fight. Officials from the Abbey had several times tried to put an end to our noisy entertainment, but they had water of a very ruddy colour thrown over them, and were so roughly used that they had to beat a hasty retreat. As the fight drew to a close the shouts increased, and the authorities, finding the noise intolerable, got one of the masters (Weare, second master) to enter the Green and stop the fight, which, as I learnt afterwards, had lasted an hour and five minutes. I believe there is an account of this fight in the old ledgers of the centre boarding-house in Little Dean's Yard. I was put into a cab and sent home, where my mother and sisters, somewhat dismayed, took charge of me, and I was made to stay for a day or two in bed.

" Within a month I got a clerkship in the Privy Council Office, and had to appear with blackened eyes and a bruised face. The Lord President, Lord Lansdowne, and two senior clerks, Harry Chester and Charles Villiers Bayley, were greatly interested in my fight, and I think helped me in getting promoted afterwards.

" More than twenty years after this event I was staying with my father in Rome, and when dining at a very large hotel dinner I recognised my old antagonist and spoke to him. He was a clergyman in poor health, and died a few years later. Some ten or twelve years later I was dining at the house of our neighbours in Earl's Court, Mr. and Mrs. Bliss, and met there Bishop Short of Adelaide, and got introduced to him and asked him about his consecration. Yes, he was consecrated with three other colonial bishops in the Abbey in 1847. I asked him if he could remember anything unusual that happened, and he at once said, ' Oh yes, there was a fight of Westminster boys, and the noise was so great that we had to complain.' He was surprised and amused when I told him that I was one of the combatants."

Nor did Westminster School monopolise Buckland's attention. It is almost needless to say that the Abbey itself occupied much of his care. " He paid the greatest attention," his son Frank writes, " to the keeping in repair of the monuments, etc., inside the Abbey, and the reparations of its external walls, applying his fund of general knowledge to the minutest details."

The Rev. W. H. Turle says :—

"I can remember how thankful we all were when Dean Buckland had the pavement in the cloister thoroughly repaired, and the gas laid on ; also he had Great Dean's Yard pavement renovated and a new gateway entrance built. The whole place was in a shamefully dilapidated condition ; the broken stonework of the bays in the cloisters was merely held together with bits of wood."

The Dean also restored all the pinnacles and buttresses on the south side of the Abbey. The monks' burying-ground —the cloister garth, the "fighting-green" of Westminster School—was turned into a stonemason's yard for several months, so great were the external repairs that were needed. Buckland carefully superintended the mason's work, whether external or internal, that was going on in the Abbey or in any other collegiate buildings in which he was interested ; he examined with his own critical and experienced eye the various kinds of cement, the blocks of building-stone, and the means adopted to repair and keep in order the regal and other monuments ; and, above all, he took special care that no faulty bits of stone were used, and that no broken pieces of monuments were thrown away.

On one occasion he received a brown-paper parcel carefully done up, containing a piece of black oak-wood about the size of a match. A letter came with it, stating that the writer, when a boy, had cut this off the coronation chair in the Abbey, and that, repenting in his old age, he returned it in the hope that it might be refitted to its old place. Buckland frequently told this story as a warning to unscrupulous collectors. At another time he received from America two small marble heads, which had been taken as a relic from Major André's tomb by some American, who, on his death-bed, had desired that they might be returned to the Abbey. With his own hands the Dean replaced these on this beautiful bas-relief. Every Sunday afternoon Buckland took his children round the Abbey, with the numerous guests who usually came to luncheon. His sharp eyes would quickly discover any fresh mutilation to any of the monuments, and he insisted on its being looked after at once. A light feather-brush which he carried in his hand served not only as a pointer, but removed the dust which always settled on the noses and outstretched fingers of the statues.

In those days it was far less common than it now is to display a reverent regard to public worship, and to take care that everything in connection with the house of God should be done with decency and order. The Dean kept a strict eye over the manner in which the services were performed, and corrected many abuses. Finding that the Abbey choristers spent their time between the services in sailing their toy boats in puddles made by the sinking of the gravestones in St. Margaret's Churchyard, or, if it were dry weather, in playing marbles on the flat slabs of the altar-tombs, he looked about for a suitable place in the precincts which could be used as a schoolroom. He found his site, opened his school, appointed an old Oxford friend to be master, and the behaviour of the choir-boys, both in and out of the Abbey, quickly improved. He made new arrangements for the greater convenience of visitors, and himself instructed the vergers in the most interesting contents of the side-chapels and other parts of the Abbey which had not hitherto been shown to the public. To the duties of the night watchman he attached great import- ance, and in Poets' Corner fixed a tell-tale clock, which registered the punctuality with which the watchman every quarter of an hour went his rounds. The cold was so intense at times in the Abbey that, as he used to say, "the fellow might go to sleep and the Abbey be burnt, as York Minster had been, from an alarm not being given in time." One of his first acts, on coming into residence, was to overhaul the fire-engine, which he found in a very crippled, useless condition. Great amusement was caused to the dwellers in the precincts by the various trials of its effici- ency, and by the exercises through which the firemen were put in Dean's Yard.

In 1848 the interior of the Abbey choir was restored. The stalls and sittings were entirely reconstructed, and, in spite of numerous objections, the Dean removed the heavy oak screens in the north and south transepts, thus adding fifteen hundred sittings to the accommodation. He took great pleasure in drawing attention to the woodwork of the stalls, many of the bosses and finials of which were carved from nature by Messrs. Ruddle, of Peterborough, showing that the modern carvers can compete in skill with their ancient brethren in the craft. The cost of these restorations amounted to over £7,000. The "marigold" window in the south transept (Poets' Corner) was filled with stained glass by Dean Buckland, as the "rose" window of the north transept had been by Dean Atterbury.

At the same time the Abbey organ was improved at the

cost of nearly £1,000 by Messrs. Hill. It was the first
cathedral organ in England to be divided into two parts
and played in the middle of the screen gallery. Mr. Hill
very well recollects Buckland asking how a thirty-two feet
pipe could lie across the aisle, which was only thirty feet
wide—a pertinent question, which Mr. Hill's father answered
by explaining that the modern sharp pitch is really a note
higher than that in vogue a hundred years ago, and reduced
the length of the pipe, so that it would just go into the
available space. It was in connection with this restoration
of the organ that Frank Buckland performed an experiment of fishing in Westminster Abbey. One of the great
open diapason pipes (wood) had become the coffin of a
deceased cat, for which the future Inspector of Fisheries set
to angle, through the top of the pipe, with a salmon hook.
In a short time he was successful and brought up " Master
Cat " in triumph.

Miss C. Fox, in her journal, speaks in the following
words of a visit to the Abbey :—

" Then to the Dean of Westminster (Dr. Buckland) in his
solemn habitation : he took us through the old Abbey, so
full of death and of life. There was solemn music going
on, in keeping with the serious Gothic architecture and the
quiet memory of the great dead. The Dean was full of
anecdote—historical, architectural, artistic, and scientific.
We got a far grander and truer notion of Westminster,
both inside and out, than we ever had before."

On Easter Day, April 23rd, 1848, the Abbey was re-
opened, after complete restoration of the choir, the congregation sitting for the first time in the transepts. On the
Continent it was a year of revolutions, and the discontent

16

at home gave serious ground for disquietude. The state
of public affairs was in the mind of Buckland, who preached
in the evening.   In his sermon he went back to the church

" built on Thorney Island, once occupied by the pagan altars
of the Roman conquerors of Britain—a site on which was
raised one of the first sanctuaries for preaching of the gospel
to our heathen forefathers, a site consecrated to God and
Christ by the piety of our Sebert, and our Offa, and our
Edgar, our Ethelred, our Alfred, and our Saxon Edward,
and nearly six centuries ago reconstructed in its actual
state of unexampled 'beauty of holiness' by our Henrys
and Edwards, in times coeval with the Crusades. . . . In this
most holy temple I and some of you have, within the last
ten months, enjoyed the privilege of witnessing the unexampled ceremony of the simultaneous consecration of
a chosen band of colonial bishops, who have gone forth
under the national sanction of the Government of this
country to preach the gospel in many of the extreme
regions of the world. . . . Never before did the compass of
Christianity circumscribe so vast a circle.

"Our modern schools of philosophy have changed their
moral phases within the present century. In the days of
our fathers and during the youth of many who are still
living, the study of philosophy was too often, and sometimes too justly, suspected to be allied to infidelity : the
study of second causes halted short of arriving at the First.
Modern professors, in carrying their researches more closely
into God's laws, by which He regulates the movement of
the material world, have been permitted to gaze more
intensely on the great source of light and life, and in every
fresh discovery they find a further and another revelation
of the infinite wisdom and power and goodness of the
Creator.

> ' Deum namque ire per omnes
> Terrasque tractusque maris cœlumque profundum.' . . .

" In the last quarter of a century the renewed spirit of

piety has planted in our island more new churches and schools than have been founded in any one or in all the centuries since the Reformation of the English Church; and already we are reaping the fruits thereof in sweet and holy experience, that 'the work of righteousness shall be peace; and the effect of righteousness quietness and assurance for ever' (Isa. xxxii. 17).

"The God of Nature has determined that moral and physical inequalities shall not only be inseparable from our humanity, but coextensive with His whole creation. He has also given compensations co-ordinate with these inequalities, working together for the conservation of all orders and degrees in that graduated scale of being which is the great law of God's providence on earth. From the mammoth to the mouse, from the eagle to the humming-bird, from the minnow to the whale, from the monarch to the man, the inhabitants of the earth and air and water form but one vast series of infinite gradations in an endless chain of inequalities of organic structure and of physical perfections : 'There are also celestial bodies, and bodies terrestrial . . . and one star differeth from another star in glory' (1 Cor. xv. 40, 41).

"So also there never was, and, while human nature remains the same, there never can be, a period in the history of human society when inequalities of worldly condition will not follow the unequal use of talents and opportunities originally the same : industry and idleness, virtue and vice, lead the same talents, with the same means and opportunities, well used or abused, to most unequal results. . . . Equality of mind or body, or of worldly condition, is as inconsistent with the order of Nature as with the moral laws of God. . . . There may be equality in poverty : equality of riches is impossible. Equality of poverty is the condition of the negro, the bushman, and the Esquimaux. Equality of wealth and property never has and never can exist, except in the imagination of wild transcendental theorists, so long as human nature shall continue to be that imperfect thing which God has placed in this world in a state of moral probation, and not of perfection. . . .

"One more last word of consolation and congratulation before we part. In the years of peril and perturbation which agitated Europe half a century ago, it was the personal character of the king of this country (King George III.) which, under Providence, was mainly instrumental to preserve us from the sanguinary revolutions which then overran the fairest part of the Continent. It is the personal character of his rightful heir and royal successor upon the throne of her ancestors which, under God's blessing, will, we trust and pray, preserve us also from the returning hurricanes of European political revolution. We know that the fervent prayer of the righteous availeth much ; and when the God of heaven beholds our most religious and gracious Queen practically affirming with the holy Joshua, ' As for me and my house, we will serve the Lord,' on her bended knees joining with her household in prayer and supplication to the King of kings and Lord of lords, we may humbly trust that the Majesty of heaven will accept the prayer of His anointed servant and minister upon earth, and in His mercy vouchsafe to hide her and the subjects of her kingdom from ' the gathering together of the froward, and from the insurrection of wicked doers.'

"England, it has been truly said, has almost always prospered under her queens. In the sacred person of our most gracious Sovereign (who within these holy walls has been anointed to rule over us), we are at this awful crisis blessed with a queen who in every relation of domestic life is a pattern of conjugal and maternal virtues, and who in her most exalted public station is the honoured exemplar of regal dignity, the object of the love and faithful service and loyal obedience of her subjects, the type and repository of mercy and clemency and supremacy, in the rule of that great united kingdom and justly balanced constitution at the head of which a gracious Providence has placed her. Blessed with such a sovereign, though the heathen may furiously rage together and the people imagine a vain thing, the throne, we trust and pray, will be exalted in righteousness and the blessing of God descend on us and our posterity." . . .

The spring of 1848 was a memorable one in London. On April 10th was the great Chartist meeting, and every preparation was made to secure the Abbey and its precincts from any rough treatment by the mob. Great alarm prevailed all over London. A hundred and fifty thousand volunteers from every walk and condition of life were sworn in as special constables. Among those who were thus sworn in was Louis Napoleon, afterwards President and subsequently Emperor. In a caricature which appeared in Paris he was represented in policemen's clothes, wielding a truncheon, with this legend : " J'ai fait plus que mon oncle, j'ai battu les Anglais dans les rues de Londres." Buckland kept his stock of these truncheons stored in the outer drawing-room for use by the Westminster specials. Every precaution was taken, and a strong guard placed in the Record Office, which then occupied the Chapter House, and in other important places. It was a remarkable feature of the day that, along the whole line of the procession, from the City to Kennington Common, the appointed rendezvous of the malcontents, scarcely a shout was raised, and only a few feeble cheers were heard.

As Feargus O'Connor was earnestly addressing the petitioners at Kennington, and entreating them not to damage their cause by any acts of violence or disorder, an eagle was seen to be soaring over their heads and flying towards Westminster ! This naturally was hailed as an excellent augury ! The bird was Frank Buckland's eagle, which had escaped that morning from the little courtyard in which it was kept. A chicken, tied by its leg to the end of a high pole, caught its keen sight towards sundown. As an

eagle never lets go its prey, the string was pulled directly
it had seized the bird, and down came the eagle. It was
easy then to throw a rug over it and cut the bird's wing.
Buckland had taken his children up to the top of the
Abbey tower in the morning to view the procession; but
the streets were empty, and as deserted by traffic as if it were
Sunday. Tired of so dull a look-out, the children descended,
and it was not till after their third journey up the innumerable stone steps leading to the tower roof that a cab was
seen driven into the Palace Yard, through a drizzling rain,
with the charter tied on to the top of it. It was, they
thought, a very poor sight after a whole day of anxious
expectation. No soldiers were to be seen ; and Buckland,
in common with the rest of London, praised the "good
tact" of the Duke of Wellington, who placed the troops in
the houses and gardens of Bridge Street and Parliament
Street, to be ready in case of emergency, but out of sight
of the mob. Mrs. Andrew Crosse tells of these troublous
times in an amusing story in the "Red-Letter Days of My
Life":—

"Other visitors there were at Broomfield [Dr. Andrew
Crosse's home] in those years, notably a party of four
distinguished men—Dr. Buckland (the then Dean of Westminster), Dr. Daubeny, Lord Playfair (then Dr. Playfair),
and Baron Liebig. These gentlemen had been inspecting
the cheese-making process of Cheddar, and, arriving at
Bridgewater, ordered a carriage and pair at the hotel,
requiring to be driven to Broomfield without loss of time.
It was the summer of 1848, the year of revolutions abroad
and Chartist alarms at home. The inn-keeper, hearing a
foreign language spoken, and learning their destination,
jumped to the conclusion that these strangers might be

plotting against Church and State, and forthwith communicated with the police, with the result that the suspicious quartet were closely watched. When the Dean of Westminster, who dearly loved a joke, heard the story subsequently, he was highly delighted with the impression they had made on the *quidnuncs* of Bridgewater."

In May 1848 Buckland and two of his daughters were attacked with typhoid, or " Westminster fever," as it was called, for it did not spread beyond the precincts. Every one was taken ill on the same day. Some workmen, in making alterations in Little Dean's Yard, accidentally opened some old drainage, and neither Buckland, who was superintending the work at the time and saw the mischief done, nor any one who was conversant with the facts, had any doubt as to the origin of the outbreak. Several deaths occurred : the unusual sound of the tolling of the Abbey bell drew attention to the fever, and caused great gloom throughout the neighbourhood. As soon as Buckland was restored to health, he lost no time in applying his scientific knowledge to the thorough cleansing and making of sewers. The system of pipe-drainage which he introduced was the first of its kind ever laid down in London. It proved completely successful. " This experiment," he says, " on the drainage and sewage of about fifteen houses and an area of about two acres affords a triumphant proof of the efficacy of draining by pipes, and the facility of dispensing entirely with cesspools and brick sewers throughout London." The experiment—for such it then was—succeeded most triumphantly. He was, therefore, deeply wounded when this outbreak of fever was ascribed to his sanitary reforms. That the charge was most unfounded is proved by the

report of the Commission employed to look into the health
of London.   The Commissioners reported as follows :—

" During the cleansing of the Westminster Abbey precincts, in the autumn of 1848, four hundred cubic yards of
foul matter had been removed from the various branches of
the ancient sewers, which were obliterated and filled up
with earth.   An entirely new system of drainage by pipes
alone was then substituted, and not a single case of failure
had been discovered by careful examinations made weekly
ever since the new pipe-drainage had been laid down."[1]

As a member of the Institution of Civil Engineers Buckland exerted himself actively in the improvement of the
supply of pure water for the Metropolis, and examined the
projects for obtaining it from the Thames and from other
rivers, and from wells sunk in the chalk.   Of all the various
plans an artesian well in the Isle of Dogs was at that time
found to yield the purest water.   On the outbreak of the
cholera, in 1848, Buckland, anxious as ever to benefit his
fellow-creatures, collected a mass of information less on the
treatment of the disease than on its prevention by care in
sanitary arrangement of the houses both of rich and poor,
and on the properties of disinfectants, with the most effica-
cious mode of applying them.   He was far ahead of his
day in sanitary science, and, like sanitary reformers of the
present time, met with endless objections to his advice to
" clean up."   In a sermon which he preached in the Abbey
on November 15th, 1849, the day of thanksgiving to God
for the removal of the cholera, he observed in allusion to
the Westminster fever, " A warning voice had not been

[1] " Report of the Metropolitan Commission of Sewers."

raised in vain, and in God's mercy we have been entirely
spared during the pestilence that has surrounded us."
This sermon on the prophet's words to Naaman, "Wash
and be clean," raised a great stir at the time. The Dean
showed how frequent and repeated were the purifications
and bodily washings " enjoined under the Mosaic Law,"
and how important the small details of cleanliness are
for us all.

" The greater number of the poor who perish," he said,.
" are the victims of the avarice and neglect of small landlords and owners of the filthy, ill-ventilated habitations in
which the poorest and most ill-fed and helpless are compelled
to dwell.   Fatal diseases are continually engendered from.
lack of adequate supplies of water, withholden from the
dwellings of the poor by the negligence of the owners, or
by the jealousy of interference by public officers or public
Boards of Health with parochial or with city authorities, or
with privileges or corporations, or with places and per-
quisites of individuals, or with established companies.   It
will be the fault of man, of the selfishness, or the folly, or
avarice of the owners of poor houses, or of the jealousy
or pride of officers and interested individuals, and it will
be the fault of Parliament also, if we do not instantly
begin to remedy these crying evils, if in two or three years
our city is not duly supplied with water.   Above all things,
cleanse your hearts, and not your garments only, and turn.
unto the Lord your God."

The offertory on this occasion was for the widows and
orphans of those who had died of the cholera in Westminster.

In medical science Dean Buckland felt a special interest.
His son Frank writes :—

" During my career at St. George's Hospital he took the

most lively interest in all that was going on there, requiring me to tell him what I had learnt at the lectures as well as the details of the more interesting cases under treatment in the wards. At the annual Hospital meeting at St. George's Hospital in 1849, at the request of the Governors, he undertook the distribution of the prizes to the students. It not unfrequently happens that these prizes are given into the hands of the successful candidates, accompanied merely by a few simple words of congratulation from the chairman ; but by those who were present on the occasion of Dr. Buckland's giving away the prizes, it will be well remembered that upon almost every subject— Anatomy, Physiology, Materia Medica, Practice of Physic, Surgery, Chemistry, etc.—he made such appropriate and apt remarks from his vast fund of general information that he seemed to throw a charm round subjects which otherwise would be dull and unentertaining to those not specially engaged in their study. . . . Amongst his numerous titles Dr. Buckland was Doctor of Medicine of the University of Bonn, which honour was conferred upon him, probably, under the idea that he was a Doctor of Medicine and not of Divinity. He was also Honorary Fellow of the Royal Medical and Chirurgical Society ; and my friend and much respected tutor in surgery, Mr. Cæsar Hawkins, as President of that Society, March 1857, thus writes of him in his obituary notice :—

" ' It is, I presume, the connection of geology with comparative anatomy and physiology, and through them with our profession, which induced the Council in 1825 to recommend Dr. Buckland as Honorary Fellow of this Society. As a comparative anatomist, Dr. Buckland and the late Mr. Clift were long consulted as the chief authorities in palæontology, by whose decision the supposed examples of exhumed bones of deceased giants were transformed into those of a modern ox or an antediluvian ichthyosaurus. Of his sagacity and readiness of conjecture, and the ingenuity with which he followed out to their consequences the relation of one fact or discovery with another in anatomy and physiology, many examples might

be given : the magnificent skeleton of the Mylodon is a
beautiful instance in which his reasoning on the probable
use of the enormous air cells between the tables of the
skull in connection with the trees it uprooted was confirmed by the safety of the real covering of the brain,
and the recovery of this large creature from enormous
fracture of the outer table, received we know not how
many thousand years ago.   It was but the necessary
tribute to his eminence in these sciences that, on his
becoming a resident of the Deanery of Westminster, Dr.
Buckland should be appointed a trustee of the British
Museum ; and also one of the trustees of the Hunterian
Museum at my own college, where he was a frequent donor
and visitor.   Among the principal of his gifts to the
museum of the Royal College of Surgeons may be
mentioned, besides numerous fossil bones, etc., the skeleton
of the now well-known gigantic bird the Dinornis or Moa,
the bones of which were sent to him by a gentleman
named Williams, whom Dr. Buckland had requested to
transmit to him any fossil bones he might find in his
missionary excursions in New Zealand ; the skeleton of
Billy the Hyena, that lived nearly a quarter of a century
under the care of the late Mr. Cross at Exeter Change,
and subsequently at the Surrey Zoological Gardens.   The
skeleton of an enormous bull-trout caught near Drayton
Manor, and presented by the late Sir Robert Peel, was
rescued from the kitchen, at Dr. Buckland's suggestion, for
a more glorious fate.[1]

    " 'Whenever lectures on any interesting subject were
given in the theatre of this most valuable, noble, and
priceless institution, Dr. Buckland was ever present, note-

---

[1] " Since becoming Inspector of Salmon Fisheries I have examined
a painting of this fish in the possession of Professor Owen at his house
in Richmond Park.   I believe it was an old salmon kelt very much out
of condition.   Fancy a Prime Minister and his learned friends sitting
down to eat an old kelt at a dinner-party !   I fear none of the *savants*
present at Drayton knew much of the salmon or of the science of
salmon culture."—See Note by Frank Buckland, Bridgewater, 4th ed.

book in hand ; but on no occasion was he a more assiduous attendant than when his friend Professor Owen gave
his admirable demonstrations on Comparative Anatomy.'
    " Dr. Buckland also applied his knowledge of human
anatomy to questions interesting to the antiquarian.   He
was present at the opening of some Saxon barrows on
Breachdown, near Canterbury, when he found ' the thick
skull, apparently, of a peasant warrior bearing marks of
a fracture received during life.' He also describes the
flattened and polished surfaces of the warrior's molar teeth,
indicating that he had eaten hard food—probably parched
peas and beans.   This fact he had frequently observed in
the teeth from the graves of ancient Britons, and also in
the teeth of modern uncivilised races of men.   On another
occasion Dr. Buckland described the claw of an eagle and
the bones of other birds found by himself in the ruins of
a Roman villa near Weymouth, and conjectures that they
were sacred birds connected with augury, or votive sacri-
fices to Esculapius ; of which we have an example in the
cock which Socrates in his dying moments commanded to
be sacrificed to that deity." [1]

The spiritual welfare of Westminster was not neglected
by the Dean.   Partly through his exertions two additional
churches were built, and, after he was himself incapacitated
by his illness, Mrs. Buckland carried on his various plans
for the alleviation of the condition of the poor.   One of
the new churches, dedicated to St. Matthew, was erected on
a site, and included a district, known as the " Devil's acre."
In Pye Street in this parish Mrs. Buckland set on foot a
coffee house, to which Her Majesty the Queen subscribed
£50, and in which many of the nobility and eminent men
of the day were interested.   The Rev. R. Malone, the
first incumbent of St. Matthew's, writes :—

---

[1] F. Buckland, Memoir to Bridgewater, 4th ed.

" Mrs. Buckland permitted me to draw up rules and to manage this novel institution. She got lecturers, and among them Frank Buckland, to give weekly lectures, and a good library was formed. It answered only too well for nearly two years, but then the police informed me it was made the meeting-place for thieves, and that they formed there schemes of burglary. On one occasion in the middle of the day I found it full of idle men, and the manager told me that directly he suggested it was not meant for a lounge for loafers, they ordered more food and kept him continually at work. I then spoke to them, and said we were anxious to make the house a comfortable and a quiet club for working men ; but that our end would be defeated if the idlers, loafers, and men who would not work crowded the rooms all day long. This was the crisis of the club, and from that day it ceased to pay, and before it failed it was thought better to close the doors."

The institution was then opened as an Industrial School for street boys. On the Committee were several barristers, among the most active of whom were the present Baron Pollock and the late Judge Bristowe. Mrs. Buckland took the greatest interest in the scheme, and helped with a large subscription. The boys were taught to make paper bags and to print ; and as they were fitted for employment, they were drafted off, and many of them became useful workmen. This coffee house was one of the first to be started in London, and was modelled upon a like refreshment place for working men in Edinburgh. Nor was it the only philanthropic scheme in which the Dean and his wife were interested.

" Mrs. Buckland," writes Mr. Malone, " gave me a small sum of money to lend out to the deserving poor, and this sum lasted a considerable time and was the means of tiding over some of the strait places in which industrious working men are sometimes placed. I well remember what a loss to my poor parish the removal of Mrs. Buckland and her family from Westminster at the death of the Dean was. It was not only her aid in money, but her practical good-sense, her kind sympathy, and her influential position, that sustained and supported."

In these charitable labours Mrs. Buckland received excellent counsel from Dean Hook. " Be thankful," he wrote, " for your successes, ignore your failures, and always be attempting something new."

CHAPTER X

ISLIP.

ISLIP, which was regarded by the family as their country home, lies on the high road between Worcester and London, seven miles from Oxford. Situated on what was formerly a great thoroughfare, it was once an active, bustling village, and is a place full of historical reminiscences. The first and most interesting of its associations with history is that it was the birthplace of Edward the Confessor, who endowed his newly founded Abbey at Westminster with his mother's birthday gift. Mr. Parker, in his " Early History of Oxford," says :—

" Eadward ' the Confessor,' elected King, was probably in Normandy at the time, and the preparations were such that he was not crowned till Easter in 1043, and then at Winchester. No traces in any charter or in any of the historians occur of his visiting Oxford. Yet one might have expected it, for it is but a few miles across the meadows on the north of Oxford to the place where he was born. This fact we do not obtain from any chronicler, but from the chance mention of it in a charter respecting a grant of land to this newly founded, or rather restored, abbey in Westminster. It runs as follows :—
" ' Eadward, King, greets Wlsy, Bishop, and Gyrth, Earl, and all my thanes in Oxnefordesyre kindly. And I would

have you to know that I have given to Christ and to Saint
Peter, unto Westminster that "cotlif" in which I was
born, by name Githslepe, and one hide at Mersce scot-free
and gafol-free, with all the things therein that thereto
belong in wood and in field, in meadow and in waters, with
church and with church-jurisdiction, as fully and as largely
and as free as it stood to myself in my hands : so also as
Elgiva Imma my mother at my first birthday gave it to me
for a provision.'"[1]

The font in which, according to tradition, Edward was
baptised, stood in the Rectory garden ; but Buckland, who
pronounced it to be fourteenth-century work, had it carefully cleaned, and presented it to a church which was being
restored in the neighbourhood.   The form of Islip is not
that of a village, but of a town ; and a "town" it is still
called, with streets branching out from an open centre
which might have been a market-place, and where a cross
once stood in front of the church.   This cross was replaced
by a lofty elm tree, which Dean Ireland had supported by
large Stonesfield slates.   The village stocks were here.   The
Rectory was built by Dr. South, Prebend of Westminster,
the famous preacher and wit, who was for thirty-eight
years Rector of the parish.   Although living occasionally
in the place, he never occupied the parsonage ; neither did
his successor, Sir R. Cope, Chaplain to the House of
Commons, who was Rector for forty years ; and it had
therefore to be restored, as, at the beginning of this century,
it was in a ruinous condition.

The Rectory,[2] and the garden, which had evidently been

---

[1] "The Early History of Oxford, 727—1100," Parker, p. 176.
[2] The roof slates are from the Stonesfield quarries, where Dr. Buckland often worked and which he frequently took his pupils to examine.

THE RECTORY, ISLIP

quarried out of solid limestone, stood on a rock elevated nearly thirty feet above the level of the river Ray, and looked upon the bridge on which Cromwell defeated the Earl of Northumberland, Lord Wilmot, and Colonel Palmer. The garden to the south was laid out in terraces, and was surrounded on all sides by walls. In this sheltered, sunny spot the Dean and Mrs. Buckland were able to cultivate a great variety of plants : stonecrop and rock cistus throve amazingly ; vines and peaches flourished ; the strawberry beds, which can be seen in the foreground, were famed far and wide ; and from some Alpine plants, brought from Switzerland, would be often picked a dish of fruit quite late in the autumn. "The roses," writes Mrs. Buckland to Faraday, in July 1849, "are now blooming, and the strawberries ripening. Our small garden is exquisitely rich in perfume." Fruit and flowers were not often to be seen in such profusion, growing side by side, as in that old seventeenth-century garden. In 1807 the garden had much good fruit planted in it by a tailor who rented the house from Sir R. Cope. Dean Vincent added a great deal more. "The best of all sorts, there is no finer fruit anywhere, and the soil is favourable," writes old Dean Vincent in a manuscript book of notes about Islip which was kept in the Parish chest, and which the present Rector, the Rev. T. Fowle, kindly lent the biographer. The village school is of Dr. South's foundation ; he managed it himself while living, and the first annual account bears date 1717, the year after his death.

Twenty-one boys, chosen from Islip in preference, then from the nearest parishes, are always to be in the school, nominated, admitted, dismissed, or chosen apprentices by the Rector or curate. Like the Rectory, which was often called the " Isle of Roads," the school stood at the entrance roads to the village, and was surrounded by roads. The lads wore the usual blue-coat costume, and were admirably taught the three R's by the schoolmaster, Mr. Chapman, a very intelligent man, whom Buckland employed to survey and measure out the allotments which he started, and also to keep record of their respective yields. Mrs. Buckland, in spite of serious remonstrances from neighbours and friends, gave the boys instruction in geography and the use of the globes, which she had made of paper and inflated, showing them at the same time on the map the homes of foreign products, and supplying specimens of the sugarcane, the tea tree, and other articles of daily use. Many amusing letters did she receive, protesting against such unnecessary teaching, which was only supposed to put foolish notions into children's heads. However, the keen interest which she awakened in their minds led ultimately to the emigration of several labourers and other families to Australia, where they have done well and have become landed proprietors. One or two have revisited their native town from time to time, but only to see their friends, and soon returned to their new possessions. Outfits were provided, and the Dean himself secured their passages, and commended them to the care of the captain. He also packed cuttings from gooseberry and currant trees in tin boxes filled with honey and soldered down to exclude the air—a mode of packing that answered well, and the emigrants had the pleasure of seeing fruit trees from Islip growing in their new gardens. All the details of their journey were carefully planned and personally superintended by the Dean. Vans met the emigrants at Paddington, and they were driven to the Deanery and hospitably entertained before going down to the docks for embarkation. " Be um aloive ? " was the general exclamation as Buckland's country friends passed the Horse Guards sentries and saw London for the first time.

The family usually spent the summer and autumn months at Islip ; and soon after taking up their residence at this pretty Rectory, schemes were set on foot for the good of the villagers. The Dean provided allotments for the labourers and directed how to lay them out. Many a summer evening was spent in chatting with and advising the labourers about the cultivation of these plots, and gaining from their practical experience much useful agricultural information. He would show them the result of the experiments he had made in a piece of ground, adjoining the allotments, which he rented for the purpose. Here, as formerly at Marsh Gibbon, experiments of one kind or another were always being made. Even the turf of Christ Church was, in former days, turned to useful account by the enthusiastic and practical farmer. Canon Jelf of Rochester, the son of Buckland's next-door neighbour as Canon of Christ Church, remembers an agricultural feat of the future Dean's. On the turf in Tom Quad, he sowed the word " guano " in this material, which had just begun to be imported from a Pacific island frequented by birds, and in due course the brilliant green grass of the letters amply testified to its efficacy as a dressing.

Some of the Westminster Prebends used to come on progress every summer to Islip—an old custom now obsolete. They met their tenants with the Dean at dinner at the Old Red Lion Inn, in the yard of which was said to have stood the Confessor's residence, with its adjoining chapel, which had been converted into a barn. The barn was still standing in the early part of this century. After the dinner was finished, and the rents were paid, these dignitaries would adjourn to the Rectory terrace for coffee, fruit, and frolic, the fruit picked by the children just before the guests arrived, "with the taste of the sun in it," as Archdeacon Jennings would say. The frolic consisted in being introduced to the various pets—the eagle, monkey, and bear, and to the tadpoles, which were kept in a pan on the terrace, and devoured one another, as did the saurians of old. A game of thimblerig followed on one occasion, played by these sedate old gentlemen with empty flower-pots. Hosts and guests were indeed a very merry company for an hour or so, after which solemn state was resumed, and the progress continued to Fencott and Mercote on Otmoor, where more Dean and Chapter property was visited and tenants interviewed.

The inhabitants of these low-lying villages suffered greatly from ague; but Islip fortunately stood just at the edge of the flat swampy stretch of land known as Otmoor. The traditional origin of the name is that a charitable lady received a promise from a great landowner that he would give her, for the benefit of the poor, as much land as she could ride round while a sheaf of oats was burning. Otmoor was like a vast lake in winter; but in spite of its apparent uselessness and swampiness, very serious riots occurred when the district was enclosed some sixty years ago. There are to be seen the remains of a fine Roman road across the moor, and the Dean would point out to the way-wardens of the fen villages how the Romans, the best road-makers in the world, made solid foundations for their streets or ways, keeping them well raised in the middle, with ditches on either side. These open drains lasted for centuries, and slabs of stone can still be plainly seen which lined the deep watercourses. After much persuasion, he succeeded in getting the roads in these marsh villages raised, the ditches kept dug out, freed from vegetable growth, and properly levelled, so that the water might flow away freely instead of becoming stagnant. This simple plan soon made its advantages felt; ague disappeared, and the health of these low-lying villages wonderfully improved. It must not be supposed that this result was gained by a few casual visits. Buckland's efforts for the health of the people were unwearied, and he never ceased to impress on his children's minds that any work undertaken, if it was to be of any value or success, must be "taken trouble with." "Never spare yourself," he said.

In 1846 the dark shadow of famine crept over the land. Not only in Ireland was the potato crop a total failure, but in England also the disease was universal. Wheat was both scarce and costly; but, till the time of scarcity came, the real importance of the potato crop had not been recognised. Buckland met the difficulty in his usual practical way. In his own household he set the example of using maize as a substitute for flour, which he only used for bread. He encouraged the villagers to make loaves of barley grown on their allotments, but could not overcome their prejudice to *black* bread. Personally Buckland enjoyed the reminiscence of his travels in Germany, when for months he subsisted on little else but barley bread and eggs. He supplied the village shops with sacks of hominy and Indian meal, which were sold for a penny or twopence a pound, and any of the " townsfolk " who liked might come to the Rectory to be taught the various ways of cooking it. Experiments were made in the manufacture of arrowroot from those tubers which were only partially affected with the disease. The whole family was set to work to grate the potatoes into pans of water ; the pulp gradually settled to the bottom, where it remained till the next day. The water was then poured off, the brown scum removed from the settlement, fresh water poured on, and, after three washings, the starch would be found snow white at the bottom of the pan. Excellent food was thus obtained, which was stored in tin boxes for the use of the poor people who had lost their winter supply of potatoes. "No waste in Nature," Dean Buckland would say.

Among other good services rendered by the Dean to Islip was the building of a cottage at the end of the large old tithe barn, one room in which was fitted up as a recreation room for the village lads. There also a night

school, then a novel institution, was held three times
a week, when some of the family were bound to be
present to provide some recreation, which often consisted of a talk about a coal, salt, or other mine—always
accompanied, if possible, by pictures or specimens, both for
illustration and " making them remember." If only a few
lads were there, the microscope was fetched. Interest was
at once keenly aroused ; and though Mr. Webb, the super-
intendent and village saddler, did his utmost to impress
upon the youths his favourite adage, " Civility costs nothing,
and gains everything," the struggles of the boys to get
" a good look through un " became somewhat difficult
to manage. Especially vigorous were the pushing and
pummelling of the spectators, when the object on view was
the last snail's tongue mounted by the Dean's youngest
daughter, or a freshly collected specimen of blight, etc.
The elder lads and men Mrs. Buckland would have up to
the Rectory, and teach them how to write letters and direct
the envelopes.

Dean Buckland always took his share of Sunday duty
with the resident curate. He left a large collection of
manuscript sermons, which for the most part are earnest,
eloquent exhortations on thoroughly practical matters.
He had, moreover, in a marked degree the faculty of
adapting his discourse to the members of his congregation, whether the learned magnates of Oxford, the simple
labourers of Islip, or the mixed audience of Westminster
Abbey.

The poor who were receiving parish relief were regularly visited ; and when bread supplied by the rates was

bad, as it often was in those days of dear wheat, the
Dean would cut off a piece of the loaf, and send it to
the guardians that they might themselves judge of its
quality. The parish doctor was a most kind old gentleman, who took almost as great an interest in Frank
Buckland's hospital progress as his father himself. On
most Saturdays, when the young medical student came
down from town, the big old Rectory kitchen would be
filled with lame, halt, and blind, sent up by the doctor
for Frank to report upon and treat in the most approved
modern way. One of his sisters had to go round with
him, and take down his directions, which she would see
carried out during the week—a training, or rather experience, that has proved of the greatest value to her during
a lifetime spent in a country parish.

The cholera was very fatal in Oxford in 1849, and there
was a great panic in all the surrounding villages, especially
in those which, like Islip, supplied the Oxford market twice
a week with dairy produce, ducks, and crayfish. Islip was
no exception to the usual insanitary condition of country
parishes, and was worse than many, owing to the constant
floods from the river Ray, a small tributary of the Cherwell,
which brought down a considerable amount of *detritus*
from the neighbouring villages on Otmoor. At this crisis
the Dean took his children with him, and visited every
cottage in his cheery, genial way, assuring the poor folk
that, if they would only keep their premises clean and
follow the advice he gave them, they need not be afraid
of cholera. Again and again he would go round the
village and see that his advice had been carried out. " If

you want a thing done well, do it yourself," he would say ; and, indeed, it was no easy matter to overcome the prejudices or lazy customs indigenous in a village community. The instinct of self-preservation, however, is as strong in the poor man as in the rich, and people can soon be brought to see the importance of keeping the well free from soakage and impurities of any kind, if only sufficient trouble and personal interest are taken to explain to less educated brethren the importance of cleanliness. "Us have never give they things a thought," was often remarked to the Dean, "but we'll clean up now a'wever" (however), "as you have showed us all about it." Happily no cholera came ; but it was very striking to find how these practical suggestions and home-to-home visits allayed the panic, which is all the more terrible when circumstances prevent people leaving a district in which an outbreak of disease has occurred. It may be added here that Buckland was the last Dean of Westminster who held the living of Islip.

In his conspicuous position as Dean of Westminster, as well as in the active administration of his retired country parish, Buckland threw his best energies into the work before him. Public honours showed the esteem which he won by his laborious and useful life.

" Perhaps of all the varied marks of honour and respect," Frank Buckland writes, " which were heaped upon him at various times by the learned societies in all parts of the world, none yielded him higher gratification than the reception, on February 12th, 1848, from the hands of Sir H. de la Bêche, of the Wollaston Medal. This is the highest mark of honour known in geological science, and

would doubtless long before have been paid to him
but for the frequency of his election to office in that
society."

In his reply to the address of the President, the Dean
used expressions such as could only be uttered by a
geologist convinced of the grand destiny of his science,
and conscious of his own right to be remembered among
the authors of discoveries whose names are inscribed on
the annals of the physical history of the globe.

" How vast are the requirements of this our own master
science, geology, with such manifold subordinates ! " said
the Dean.   " What a mighty miracle is the earth which it
is our province and privilege to investigate !   How highly
calculated is the study of its structure to awaken many
of the most exalted feelings of our spiritual nature—
feelings kindred to those of which original first discoverers
of the laws and principles that govern the material world
must occasionally be conscious—feelings of grateful and
humble admiration of the Great Author of all created
things, which exalt us in the scale of beings, and which I
once experienced when, standing on the highest summit
of the Mendip Hills, at the close of an elaborate investigation of the structure of the surrounding country, I
recollected that I was the first individual of the human
race to whom it had been permitted to unravel the
structure and record the history of that portion of the
works of God that lay within the horizon then around
me.

" It has been the high privilege of our time which our
successors cannot enjoy to be the pioneers of a great and
comprehensive master science ;   and wherever we have
pushed forward our original discoveries, these discoveries
will have indelibly inscribed our names on the annals of
the physical history of the globe.   We have established
landmarks and fixed physical and chronological horizons,

which must endure so long as men regard the structure, and contents, and physical history of the earth which God has given to the children of men.

"Geological knowledge, *i.e.* the knowledge of the rich ingredients with which God has stored the earth beforehand, when He created it for the then future use and comfort of man, must fill the mind of every one who acquires this knowledge, with feelings of the highest admiration, the deepest gratitude, and the most profound humility. The more our knowledge increases, of the infinity of the wisdom and goodness of the Creator, greater and greater becomes the consciousness of our own comparative ignorance and insignificance. The sciolist alone is proud; the philosopher is humble, and duly conscious of the comparative littleness of his most extended knowledge. We may be gratified by our discoveries, and by the recognition of the value of our labours by our fellow-men. We may and ought to be gratified, but we are not made proud; we feel that pride was not made for man; we learn the lesson of humility; increasing more and more continually, as our knowledge of the works of God becomes more and more expanded; and to those who have laboured diligently and successfully in their calling as investigators of the wonders of creation, it is permitted to hope that they have done good in their generation, and that their labour has not been in vain."

The presentation of the medal was almost the last important occasion on which Buckland appeared in public. The words already quoted from his speech—which, as Sir Roderick Murchison writes of his old friend, "embody a humble confession of the comparative littleness and incompleteness of all human knowledge"—were but "too prophetic of the approaching close of his own valuable and honourable career."

The first few months of the Dean's mysterious illness

were spent at the Deanery. The best medical opinions were consulted in vain. The cause of the illness baffled the highest skill, but to the last it was hoped that the malady might disappear as mysteriously as it had come. Acting on the advice of the first doctors of the day, Buckland continued to hold his Deanery, the duties of the office being discharged by the Sub-Dean, Lord John Thynne. " Lord John," writes Mrs. Buckland to Faraday, " has carried on the business of the Chapter for my poor husband. You may judge how deeply I feel indebted to him." But science proved unavailing. Nothing relieved the apathetic gloom and depression which gradually settled down upon this gifted man. As the symptoms became worse, his doctors recommended the quiet and fresh air of Islip, since medical remedies proved of no avail against the peculiar and apparently unprecedented malady of which he was the victim.[1] The sight of the garden and his favourite allotments seemed to cheer him for a time, but the terrible weakness, torpor, and loss of flesh rapidly

---

[1] Frank Buckland writes: " In a medical point of view Dr. Buckland's illness is at once most interesting and important. The best medical opinions could decide only as to the symptoms and treatment of the malady; the real cause of the cerebral disturbance, and consequent mental suffering, was never suspected, and was ascertained only after death. No symptom of it, strange to say, was ever exhibited in life, and even if it had been, medical aid would have been unavailing. Those who made the examination ascertained that the brain itself was perfectly healthy in every respect; but the portion of the base of the skull upon which the brain rested, together with the two upper vertebræ of the neck, were found to be in an advanced state of caries, or decay. The irritation, therefore, communicated by this diseased state of the bones above was quite sufficient cause to give

increased. Sir Roderick Murchison would often visit his well-beloved friend, and endeavour to interest him in his old pursuits ; but nothing roused him. The "*Leisure Hour*," Frank Buckland says, "was the only publication my dear father would read during his illness, and the volumes were always on the table ; he would look at nothing else, save the Bible."

During the Dean's illness Mrs. Buckland and her daughters lived chiefly at Islip, within reach of all the old Oxford friends, and constantly visited by Murchison, Owen, Harcourt, Conybeare, and others. In a letter of invitation to Faraday she writes :—

"This place has no fine scenery, but I think you and Mrs. Faraday will like the village quiet and the sunny terrace. I shall not attempt to lionise you, far less to make lions of you ; but you shall have the sincere welcome which I have so much pleasure in offering to my poor husband's valued friends. He is, as usual, well, and not unhappy when left in perfect repose—a strange contrast to his former existence ! "

rise to all symptoms ; this irritation being considerably augmented by continuous and severe 'exercise of the brain in thought.' My parents, when travelling to a scientific meeting in Berlin, met with a severe accident ; the diligence was overturned, my father fell from the top and was stunned, and unable to render any assistance to my mother, who received a deep cut on her frontal bone. Professor Ehrenberg, who fortunately was with them, attended to the injuries they there received, which proved the ultimate cause of death in both. Dr. Buckland's vertebræ were injured, and a bony tumour was discovered to have formed at the back of the cut on Mrs. Buckland's frontal bone, which, for the last two years of her life, occasioned attacks of unconsciousness, in one of which she died."

Sir Roderick Murchison was constantly with the Bucklands—ever ready with sympathy and advice, the very *beau idéal* of a friend. The following letter is an evidence of the regard he had for both the Dean and his wife :—

" 16, BELGRAVE SQUARE,
"*July 5th,* 1854.

" MY DEAR MRS. BUCKLAND,—If you had been in town, it was my intention to have begged your acceptance of my 'Siluria'; and if you are now at Islip, will you tell me whether and how to send it to you?

" You will be the *only* lady to whom a copy is sent, and I make this special exception out of sincere regard for yourself and gratitude to your husband, who helped on the old soldier to make his way as a geologist. I have in a prelude to the work explained how Dr. Buckland was the first person who incited me to examine *the very tract* in which I opened out the mine that proved so rich and instructive.

" I well recollect our pleasant visit to you in 1831 on our way to Wales, and when I was looking out for some entirely *fresh pastures* and exercises for my restless mind. Alas! what changes since ; among these none grieved me more than the visitation with which you and your family were afflicted.

" My book necessarily deals little with the subjects in which my eminent friend most distinguished himself, but the two or three allusions made to him will, I trust, gratify you. Lyell, albeit my last chapter pokes him very hard, has complimented me much on the work, and particularly for the manner in which I have handled the Cambrian shadows which have melted away before the labours of so many good men : none of them certainly were paid or bribed by me.

" My case is simply that of truth, as old Lonsdale writes, and I cannot be put aside.

" My preface and the three lines of dedication to De la Bêche, with a map literally made from the Government surveys, prevent all further dispute.

" Ever your sincere friend,

" RODERICK MURCHISON."

Dean Buckland died August 14th, 1856, at the advanced age of seventy-three. He was buried at the west end of the churchyard at Islip. The spot, which was selected by himself, lay beside the terrace gravel walk with its row of elms. From it he had often taken his children to gaze on the beautiful sunsets lighting up the wide stretch of low, level landscape, with Kidlington spire " pointing up to heaven like a needle," he would say, in the golden haze which melted into the purple of the Witham Hills on the distant horizon. Curiously enough, his grave had to be hewn out of the solid limestone, and blasting powder was used in considerable quantities to excavate the rock. Mrs. Buckland restored the chancel, at the cost of £500, in memory of the Dean, and replaced Dr. South's very ugly east window by new stone tracery, which, after her death at St. Leonards in the following year, November 1857, her children filled with stained glass, to the memory of both their parents.

With the permission of the Dean and Chapter his children placed a monumental bust in the south aisle of Westminster Abbey, near the door leading to the cloisters. The following is the inscription on the plinth, written by the Rev. the Sub-Dean, Lord John Thynne :—

IN MEMORY OF
THE VERY REV. WILLIAM BUCKLAND, D.D., F.R.S.,
DEAN OF WESTMINSTER,
AND OF THE MOST HONOURABLE ORDER OF THE BATH,
FORMERLY CANON OF CHRIST CHURCH, OXFORD;
TRUSTEE OF THE BRITISH MUSEUM;
FIRST PROFESSOR OF GEOLOGY AND MINERALOGY
IN THE UNIVERSITY OF OXFORD;
FOUNDER OF THE MUSEUM OF GEOLOGY WHICH HE BEQUEATHED
TO THAT UNIVERSITY.

Endued with superior Intellect,
He applied the Powers of His Mind
To the Honour and Glory of God,
The advancement of Science,
And the welfare of Mankind.

BORN MARCH 12, 1784: DIED AUGUST 14, 1856. AGED 73.

" For the Lord giveth wisdom : out of His mouth cometh knowledge and understanding."—PROVERBS ii. 6.

ERECTED BY HIS CHILDREN.

After Buckland's death his widow and children went to Brighton for a few months, to look for a new home, which was found at East Ascent, St. Leonards. Although in feeble health, Mrs. Buckland continued to work at the microscope, with her daughter Caroline, upon marine zoophytes and sponges, as she had done at Islip on fresh-water animalculæ and plants. Dr. Bowerbank, F.R.S., her valued friend, in a letter to Mr. Henry Lee, thus kindly expresses his appreciation of Mrs. Buckland :—

" I can assure you I feel in no small degree indebted to my late kind friend Mrs. Buckland, who assisted me with rare specimens of sponges collected by her at Guernsey and Sark, which I certainly should not have had to describe in my work on those subjects without her aid. During her residence at St. Leonards I spent many very pleasant hours in her society, and she was an earnest and acute observer to the last. On November 29th, 1857, the day preceding her decease, I spent the morning with her in microscopical investigations, and when I took leave of her at two o'clock she made me promise to come on the Monday following to renew our observations ; but on the evening of the day following our meeting she was no more, to the deep regret of all who knew and appreciated her talents and her amiability."

Mrs. Buckland is buried in the same grave with her husband ; and their son Frank, in his fourth edition of the Bridgewater Treatise, published in 1869, writes :—

"A simple but lasting monument of polished Aberdeen granite records the last resting-place of as good a man and wife as ever did their duty towards God and towards their fellow-creatures."

THE END.

18

APPENDIX.

A LIST OF DR. BUCKLAND'S APPOINTMENTS AND LITERARY TITLES.

Dean of Westminster, 1845.

Canon of Christ Church, Oxford, 1825.

Professor of Geology and Mineralogy in the University of Oxford, 1818.

Fellow of the Royal Society, 1818 : Copley Medal, 1822 ; Member of the Council from 1827 to 1849 ; Vice-president, 1832—33.

Trustee of the British Museum, 1847.

Fellow of the Geological Society : (twice) President, 1824—25, 1840—41 ; Wollaston Medal, 1848.

President of British Association, 1832.

Fellow of the Linnæan Society, 1821.

Hon. Fellow of the Medico-Chirurgical Society, London, 1836.

Fellow of the Geographical Society.

Royal Institute of British Architects, Hon. Member, 1846.

Fellow of the Zoological Society.

British Archæological Society, Member.

The Naval and Military Library and Museum, Whitehall, Hon. Member.

The Institution of Civil Engineers, Hon. Member, 1842.

Royal Agricultural Society of Great Britain, Member.

Ashmolean Society, Oxford, Member.

Shropshire and North Wales Natural History and Antiquarian Society, Hon. Member, 1823.

The Philosophical Society of Bristol, Hon. Member, 1823.

The Worcestershire Natural History Society, Hon. Member.

The Cambrian Society of Geology, etc., Swansea, Hon. Member, 1824.

The Northern Institution of Science and Literature of Inverness, Hon. Member, 1825.

The Natural History Society of the Counties of Northumberland, Durham, and Newcastle, Hon. Member, 1829.

The Whitby Literary and Philosophical Society, Hon. Member, 1831.

The Bedford Natural History Society, Member, 1832.

The Leeds Philosophical and Literary Society, Hon. Member, 1835.

The Warwickshire Natural History and Archæological Society, Hon. Member, 1837.

The Geological and Polytechnic Society of the West Riding of Yorkshire, Hon. Member, 1838.

The Birmingham Philosophical Society, Hon. Member, 1838.

The Literary and Philosophical Society of St. Andrews, Hon. Member, 1838.

The Literary and Philosophical Society of Manchester, Hon. Member, 1843.

Member of Dr. Johnson's Club, 1829.

Hon. Member of Tasmanian Society.

AMERICAN.

American Geological Society, at New Haven, Connecticut, Hon. Member, 1822.

The American Academy of Arts and Sciences, Massachusetts, Fellow, 1825.

The New York Lyceum of Natural History, Hon. Member, 1828.

The Literary and Historical Society of Quebec, Fellow, 1834.

The Geological Society of Pennsylvania, Corresponding Member, 1834.

The Boston Society of Natural History, Hon. Member, 1837.

The National Institute for the Promotion of Science, Washington, Corresponding Member, 1844.

FOREIGN.

The Imperial Societies of Mineralogy and Natural History at St. Petersburg and Moscow, Member, 1818.

Muséum d'Histoire Naturelle, au Jardin du Roi, Corresponding Member, 1821.

Société Géologique de France, Member, 1821.

Sociéta Reale Borbonica Accademia delle Scienze, Naples, Corresponding Member, 1821.

Cæsare Leopoldino Carolinæ Academiæ Naturæ Curiosarum Bonenæ, Fellow, 1822.

Die Naturforschende Gesellschaft zu Halle, Corresponding Member, 1823.

Gesellschaft des Vaterlandeschen Museum in Bohmen, Diploma, 1824.

Academia Scientiarum Instituti Bonaniensis, Fellow, 1833.

Die Naturforschende Gesellschaft zu Frankfurt-am-Main, Corresponding Member, 1833.

L'Accademia delle Scienze di Bologna, Diploma, 1833.

The Asiatic Society, Bengal, Hon. Member, 1834.

Physiographiska Sällskapet I. Lund (Sweden), Corresponding Member, 1837.

University of Bonn, Diploma Doctoris, 1838.

Institut de France, Académie Royale des Sciences, Corresponding Member, 1839.

Société Française de Statistique Universelle de Paris, Member, 1839.

Société d'Agriculture et des Arts de Boulogne-sur-Mer, Hon. Member, 1839.

Institut des Provinces de France, Corresponding Member, 1841.

Societas Artium et Doctrinarum apud Theno Trajectinos, Diploma, 1842.

Die Naturevissenschaftliche Gesellschaft in Dresden, Diploma, 1845.

L'Accademia Valdarno, Corresponding Member, 1846.

University of Prague, Diploma Doctoris, 1848.

PUBLISHED WRITINGS OF PROFESSOR WILLIAM
BUCKLAND.

1.—Description of an Insulated Group of Rocks of Slate and
   Greenstone in Cumberland and Westmoreland. Geological Society Transactions, IV., 1817.
2.—Description of a Series of Specimens from the Plastic Clay
   near Reading, Berks. Geol. Soc. Trans., 1817.
3.—Description of the Paramoudra, a singular fossil body that is
   found in the chalk of the North of Ireland. Geol.
   Soc. Trans., 1817.
4.—Notice on the Geological Structure of a Part of the Island
   of Madagascar. Geol. Soc. Trans., V., 1821, and Journ.
   de Physique, XCIII., 1821.
5.—Description of the Quartz Rock of the Lickey Hill in
   Worcester, etc. Geol. Soc. Trans., V., 1821.
6.—Instructions for Conducting Geological Investigations and
   Collecting Specimens. Sillimans' Journal, III., 1821.
7.—On the Structure of the Alps, and their Relation to the
   Secondary and Transition Rocks of England. Thomson,
   Ann. Phil. T., 1821.
8.—Account of an Assemblage of Fossil Teeth and Bones of
   elephant, rhinoceros, hippopotamus, bear, tiger, hyena,
   and sixteen other animals, discovered in a cave at
   Kirkdale, Yorkshire, in the year 1821. Phil. Trans.,
   1822, and various other Journals in 1822 and 1823.
9.—On the Excavation of Valleys by Diluvial Action, as illustrated
   by a succession of valleys which intersect the South
   Coast of Dorset and Devon. Geol. Soc. Trans., 1824.
10.—Observations on the South-Western Coal District of England.
   Geol. Soc. Trans., I., 1824.

11.—Notice on the Megalosaurus, or Great Fossil Lizard of Stonesfield. Geol. Soc. Trans., I., 1824.

12.—Reply to some Observations in Dr. Fitton's Remarks on the Distribution of the British Animals. Edinb. Phil. Jour., XII., 1825.

13.—On the Discovery of the Anoplotherium Commune in the Isle of Wight. Thomson, Ann. Phil., X., 1825.

14.—Relation d'une decouverte récente d'os fossiles faite dans la partie orientale de la France. Ann. Sci. Nat., X., 1827.

15.—On the Interior of the Dens of Living Hyenas. Edinb. New. Phil. Journal, XIV., 1827.

16.—Observations on the Bones of Hyenas and other Animals in the Cavern of Lunel. Edinb. Journal Sci., VI., 1827.

17.—Notes sur les traces de Tortues observées dans le Gres rouge. Ann. Sci. Nat., XIII., 1828.

18.—On the Formation of the Valley of Kingsclere and other Valleys by the Elevation of the Strata that enclose them, etc. Geol. Soc. Trans., 1829.

19.—Geological Account of a Series of Animal and Vegetable Remains and of Rock collected by J. Crawfurd, Esq., on a Voyage up the Irawadi to Ava in 1826—27. Geol. Soc. Trans., 1829; Ann. Sci. Nat., XIV., 1828.

20.—On the Cycadeoideæ, a Family of Fossil Plants found in the Oolitic Quarries of Paviland. Geol. Soc. Trans., II., 1829.

21.—Supplementary Remarks on the Supposed Power of the Waters of the Irawadi to convert Wood to Stone. Geol. Soc. Trans., II., 1829.

22.—Letter on the Discovery of Coprolites in North America. Phil. Mag., VIII., 1830.

23.—On the Fossil Remains of the Megatherium recently imported into England from South America. Brit. Assoc. Rep., I., 1832.

24.—Appendix to Mr. de la Bêche's paper on the Geology of Nice. Geol. Soc. Proc., I., 1834.

25.—On the Discovery of a New Species of Pterodactyle, and
    also of Fæces of the Ichthyosaurus and of a Black
    Substance resembling Sepia, in the Lias at Lyme Regis.
    Geol. Soc. Proc., I., 1834.

26.—On the Vitality of Toads enclosed in Stone and Wood.
    Zool. Jour., V., 1832—34.

27.—On the Occurrence of Agates in Dolomitic Strata of the
    New Red Sandstone Formation in the Mendip Hills.
    Geol. Soc. Proc., I., 1834.

28.—On the Discovery of Fossil Bones of the Iguanodon in the
    Iron Sand of the Wealden Formation in the Isle of
    Wight, and in the Isle of Purbeck. Geol. Soc. Proc.,
    I., 1834, and Geol. Soc. Trans., III., 1835.

29.—Observations on the Secondary Formations between Nice
    and the Col di Tende. Geol. Soc. Trans., III.,
    1835.

30.—Uber den Bau und die mechanische Kraft des Unterkiefers
    des Dinotherium. Leonard u. Breun, N. Jahrb., 1835.

31.—Notiz uber die hydraulische wirkung des Sephons bei
    den Nautilear Ammoniter, u. anderen Polythalamien.
    Leonard u. Breun, N. Jahrb., 1835.

32.—On the Fossil Beaks of four Extinct Species of Fishes,
    referable to the genus Chimæra, that occur in the
    Oolitic and Cretaceous Formations of England. Phil.
    Mag., VIII., 1836.

33.—Bernerkungen uber das genus Belemnosepia und uber den
    fossilen Dinten-sack in dem vorderen Kegel der Belem-
    niten. Leonard u. Breun, N. Jahrb., 1836.

34.—On the Adaptation of the Sloths to their peculiar Mode of
    Life. Linn. Soc. Trans., XVII., 1837.

35.—Account of the Fossil Footsteps of the Cheirotherium, etc.,
    in the stone quarries of Storeton Hill, near Liverpool.
    Brit. Assoc. Rep., VII., 1838.

36.—Notice of a Newly Discovered Gigantic Reptile. Geol. Soc.
    Proc., II., 1838.

37.—On the Occurrence of Silicified Trunks of Trees in the New Red Sandstone at Allesley. Geol. Soc. Proc., II., 1838.

38.—On the Discovery of Fossil Fishes in the Bagshot Sands. Geol. Soc. Proc., II., 1838.

39.—On the Discovery of a Fossil Wing of a Neuropterous Insect in Stonesfield Slate. Geol. Soc. Proc., II., 1838.

40.—On the Fossil Fishes in the Bagshot Sand at Goldworth Hill, four miles north of Guildford. Phil. Mag., XIII., 1838.

41.—On the Action of Acidulated Waters on the Surface of the Chalk near Gravesend. Brit. Assoc. Rep., 1839.

42.—On the Agency of Animalcules in the Formation of Limestone. Ashmolean Soc. Proc., XVII., 1840.

43.—On Modes of Locomotion in Fishes. Ashmolean Soc. Proc., II., 1843—52.

44.—On Recent and Fossil Semicircular Cavities caused by air-bubbles in the surface of the soft clay, and resembling impressions of rain-drops. Brit. Assoc. Rep., 1842.

45.—On the Former Existence of Glaciers in Scotland and in the North of England. Edinb. New Phil. Journ., XXX., 1841 ; Geol. Soc. Proc., III., 1842.

46.—On the Agency of Land-Snails in corroding and making deep excavations in compact Limestone Rocks. Geol. Soc. Proc., III., 1842.

47.—On the Glacio-diluvial Phenomena in Snowdonia and the adjacent parts of North Wales. Geol. Soc. Proc., III., 1842.

48.—On Artesian Wells. Edinb. New Phil. Journ., XXXVII., 1844.

49.—On the Mechanical Action of Animals on hard and soft Substances during the Process of Stratification. Brit. Assoc. Rep., 1845.

50.—On Ichthyopatolites, or Petrified Trackways of Ambulatory Fishes upon Sandstone of the Coal Formation. Geol. Soc. Proc., IV., 1843.

19

51.—On the Occurrence of Nodules (called Petrified Potatoes) found on the shores of Lough Neagh, Ireland. Geol. Soc. Journ., II., 1846.

52.—On the causes of the general presence of Phosphates in the strata of the earth, and in all fertile soils ; with observations on Pseudo-coprolites, and on the possibility of converting the contents of Sewers and Cesspools into Manure. Agric. Soc. Journ., X., 1849.

Buckland, William, and Conybeare.

Observations on the South-West Coal District of England. Geol. Soc. Trans., I., 1824.

Buckland, William, and de la Bêche.

On the Geology of Weymouth and the Adjacent Parts of the Coast of Dorsetshire. Geol. Soc. Proc., I., 1834.

Buckland, William, and Milne.

Report of the Committee appointed in 1842 for registering the Shocks of Earthquakes, and, making such Meteorological Observations as may to them appear desirable. Brit. Assoc. Rep., 1843.

The above list is condensed from the Royal Society's " Catalogue of Scientific Papers published between 1800 and 1863," but it does not include :—

Reliquiæ Diluvianæ ; or, Observations on Organic Remains Attesting the Action of an Universal Deluge. London, 1823.

Vindiciæ Geologicæ ; or, the Connection of Geology with Religion explained in an Inaugural Lecture delivered before the

University of Oxford, May 15th, 1819, on the Endowment of a Readership in Geology by H.R.H. the Prince Regent. Oxford, 1820.

On Geology and Mineralogy considered with reference to Natural Theology. Two Vols. London, 1836.

Addresses to the British Association, 1832 and 1833.

Addresses to the Geological Society, 1840 and 1841.

INDEX.

Printed by Hazell, Watson, & Viney, Ld., London and Aylesbury.

Printed in the United States
By Bookmasters